Junior Astronomy Notebooking Journal

for

Exploring Creation with Astronomy

by
Jeannie Fulbright

Junior Astronomy Notebooking Journal

Published by
Apologia Educational Ministries, Inc.
1106 Meridian Plaza, Suite 220
Anderson, IN 46016

www.apologia.com

Copyright © 2011 Jeannie Fulbright. All rights reserved.

Manufactured in the United States of America
First Printing: July 2011

ISBN: 978-1-935495-59-8

Printed by Courier Printing, Kendallville, IN

Cover Design by Kim Williams
Cover photos courtesy NASA/JPL/Caltech

All Biblical Quotations are from the New American Standard Bible, English Standard Version, New International Version, New King James Version.

Photo Credits

Images by Rebecca Purifoy: 10, 11, 20, 21, 36, 37, 51, 52, 65, 66, 79, 80, 91, 92, 103, 104, 114, 115, 124, 125, 135, 136, 147, 148, 161, 162, 173, 174

NASA.gov: Cover, pp. 9, 16, 38, 39, 48, 53, 54, 59, 67, 68, 73, 76, 81, 82, 86, 87, 93, 94, 98, 101, 116, 120, 121, 129, 137, 144, 149, 158, 167, 176, 184, 189, 193, 194, 195, 197, 198, 201, A5, A7, A9, A17, A25, A29, A33, A35, A41, A45, A47, A49, A51, A53, A59, A61

Crestock Images: 69, 175, A27 (scale), A55 (astronaut)

Brad Fitzpatrick Images: 22, 23, 24, A11, A21, A23, A37

Angie Coleman of Artistic Energy: A19

All other images licensed from Jupiter Images.

Font Credits

Fonts used with permission/license from:

David Rakowski - Starburst
P22 Foundry, Inc. - LTC Fourneir
Font Diner: Starburst Lanes Twinkle, Bingo, Fontdinerdotcom Sparkly, Black Night, Spacearella, Hothead, Mirage Zanzibar
Educational Fontware, Inc - HWT
John Stracke - Rockets
Brian Ether - Orbitronio
My Fonts.com: Astype - Alea, Intellecta Design - Ulma, Bailarina, Worm, Gavinha, Half Flower, Wundes - Sprouts, Wiescher Desgin - Barracuda, Vivian, TypeSetit - FleurDeLeah, Bomparte - Black Swan, Scriptorium - Morris Inhas, Mandragora, Nick's Fonts - HarvestMoon, Otto Maurer - Hotrod, Arttypes - Maria Belle, Studiocharlie - Super Starlike, HiH Retrofonts - Waltari, Gradl, Aah Yes - Starbell Two, FontHaus - Novella, BitStream, Inc

Note from the Author

Welcome to the wonderful adventure in learning called "Notebooking." This notebooking journal correlates with Apologia's *Exploring Creation with Astronomy*, by Jeannie Fulbright. The activities in this journal provide everything your child needs to complete the assignments in *Exploring Creation with Astronomy* and more. It will serve as your child's individual notebook. You only need to provide scissors, glue, colored pencils, a stapler and some brass fasteners.

The concept of notebooking is not a new one. In fact, keeping notebooks was the primary way the learned men of our past educated themselves, from Leonardo Da Vinci and Christopher Columbus to George Washington, John Quincy Adams and Meriwether Lewis. These men and many others of their time were avid notebookers. As we know, they were also much more advanced in their knowledge—even as teens—than we are today. George Washington was a licensed surveyor during his teenage years, and John Quincy Adams graduated from law school at age 17.

It would be wise for us to emulate the methods of education of these great men, rather than the failing methods used in our schools today. Common modern methods, namely fill-in-the-blank and matching worksheets, do not fully engage the student's mind. Studies show that we remember only 5% of what we hear, 50% of what we see and hear and 90% of what we see, hear and do. When we participate in activities that correspond with learning, we increase our retention exponentially. This is exactly what the Junior Astronomy Notebooking Journal is designed to do—offer engaging learning activities to increase your student's retention.

The National Center for Educational Statistics shows us that American school children, by twelfth grade, rank at the bottom of international assessments, and do not even know 50% of what students in top ranked countries know. As home educators, we have the opportunity to discard methods that are detrimental and ineffective, and adopt the methods which will genuinely educate our children.

In addition to academic achievement, notebooking offers many benefits to students and parents. For students, it provides an opportunity to uniquely express themselves as they learn. It also provides a treasured memento of educational endeavors. For parents, it is a record of the year's studies and can easily be transferred to a portfolio if needed.

This journal will make notebooking easier for both you and your student by supplying a plethora of templates, hands-on crafts and projects, additional experiment ideas, and many activities that will engage your student in learning. It will prove invaluable in helping students create a wonderful keepsake of all they learned in astronomy. Remember that everything in this notebooking journal is optional. Because it will serve as your student's own unique notebook, you may customize it by simply tearing out the activity pages that you choose not to use. You, as the teacher, will decide what truly benefits your student's learning experience, encourages a love for learning and builds his confidence in science. Every child is different, learns differently and will respond differently to the array of activities provided here. Use discernment in how many of the activities and assignments you use with your child. Your goal is not to complete every activity, but to make learning a joy.

However, as a seasoned home educator, let me encourage you not to attempt to do every single activity in this notebooking journal. Choose the projects and activities that will be enjoyable and inspire a love of learning. If something is a drudgery, it will not serve to increase your student's retention, but will only discourage their enjoyment of science—resulting in an unmotivated learner.

It is my hope and prayer that you and your students will benefit from your studies this year, growing closer to God as you learn of His creation, and finding joy in the learning process.

Warmly,

Jeannie Fulbright

Table of Contents

Descriptions and Instructions for Each Page in Journal 6
Daily Schedule/ Reading Guide 7
Journal Owner Cover Page 9

Lesson 1 What is Astronomy? 10
    Coloring Pages About Astronomy 10
    Fascinating Facts About Astronomy 12
    Notebooking Activity: Mnemonic 13
    Scripture Copywork 14
    Vocabulary Crossword 16
    Project Page 17
    Minibook Paste Page 18
    Take It Further 19

Lesson 2 The Sun 20
    Coloring Pages About the Sun 20
    Fascinating Facts About the Sun 22
    Mid Lesson Activity: Write a Speech 24
    Notebooking Activity: Sun Collage 25
    Scripture Copywork 26
    Vocabulary Lift The Flap 29
    Project Page 32
    Minibook Paste Page 33
    Take It Further 34

Lesson 3 Mercury 36
    Coloring Pages About Mercury 36
    Fascinating Facts About Mercury 38
    Scripture Copywork 40
    Vocabulary Puzzle Game 43
    Project Page 48
    Minibook Paste Page 49
    Take It Further 50

Lesson 4 Venus 51
    Coloring Pages About Venus 51
    Fascinating Facts About Venus 53
    Notebooking Activity: Comic Strip 55
    Scripture Copywork 56
    Vocabulary Story 58
    Project Page 59
    Notebooking Project: How Radar is Used 60
    Minibook Paste Page 63
    Take It Further 64

Lesson 5 Earth 65
    Coloring Pages About Earth 65
    Fascinating Facts About Earth 67
    Notebooking Activity: Advertisement 69
    Scripture Copywork 70
    Vocabulary Lift The Flap 73
    Project Page 76
    Minibook Paste Page 77
    Take It Further 78

Lesson 6 The Moon 79
    Coloring Pages About the Moon 79
    Fascinating Facts About the Moon 81
    Notebooking Activity: Chart the Moon 83
    Scripture Copywork 84
    Moon Phases 86
    Project Page 87
    Minibook Paste Page 88
    Take It Further 89

Lesson 7 Mars 91
    Coloring Pages About Mars 91
    Fascinating Facts About Mars 93
    Notebooking Activity: Going to Mars 95
    Scripture Copywork 96
    Vocabulary Crossword 98
    Project Page 99
    Minibook Paste Page 100
    Take It Further 101

Lesson 8 Space Rocks 103
    Coloring Pages About Space Rocks 103
    Fascinating Facts About Space Rocks 105
    Notebooking Activity: Comets, Meteors, Asteroids 106
    Notebooking Activity: Meteor Showers 107
    Scripture Copywork 108
    Vocabulary Story 110
    Project Page 111
    Minibook Paste Page 112
    Take It Further 113

Lesson 9 Jupiter 114
    Coloring Pages About Jupiter 114
    Fascinating Facts About Jupiter 116
    Notebooking Activity: Make a Newspaper 117
    Scripture Copywork 118
    Jupiter Vocabulary 120
    Project Page 121
    Minibook Paste Page 122
    Take It Further 123

Lesson 10 Saturn 124
    Coloring Pages About Saturn 124
    Fascinating Facts About Saturn 126
    Notebooking Activity: Make a Venn Diagram 128
    Vocabulary Crossword 129
    Scripture Copywork 130
    Project Page 132
    Minibook Paste Page 133
    Take It Further 134

Table of Contents

Lesson 11 Uranus and Neptune 135
    Coloring Pages About Uranus and Neptune 135
    Fascinating Facts About Uranus and Neptune 137
    Notebooking Activity: The Discovery of Uranus 138
    Neptune and Uranus Match Up 139
    Scripture Copywork 142
    Project Page 144
    Minibook Paste Page 145
    Take It Further 146

Lesson 12 Pluto and the Kuiper Belt 147
    Coloring Pages About Pluto and the Kuiper Belt 147
    Fascinating Facts About Pluto and the Kuiper Belt 149
    Notebooking Activity: Pluto Debate 150
    Vocabulary Puzzle Game 151
    Scripture Copywork 156
    Project Page 158
    Minibook Paste Page 159
    Take It Further 160

Lesson 13 Stars and Galaxies 161
    Coloring Pages About Stars and Galaxies 161
    Fascinating Facts About Stars and Galaxies 163
    Notebooking Activity: Constellations in my Area 165
    Notebooking Activity: Constellations I found 166
    Stars and Galaxies Identification 167
    Scripture Copywork 168
    Project Page 170
    Minibook Paste Page 171
    Take It Further 172

Lesson 14 Space Travel 173
    Coloring Pages About Space Travel 173
    Fascinating Facts About Space Travel 175
    Notebooking Activity: Let's Visit the Planets 176
    Vocabulary Puzzle Game 177
    Scripture Copywork 182
    Project Page 184
    Minibook Paste Page 185
    Minibook Paste Page 186
    Take It Further 187

Answers to the Vocabulary Activities 188
Field Trip Sheets 203
Minibooks .... Appendix

Junior Astronomy Notebooking Journal

Below are descriptions of a suggested schedule and the activities included in this notebooking journal.

Suggested Schedule

A suggested schedule for reading the *Exploring Creation with Astronomy* text and completing the activities contained in the book and in this journal has ben provided. Please do not feel the need to complete every activity or assignment. Use the schedule as a guide, in a way that best suits your family.

Coloring Pages, Notebooking Assignments, Activites and Projects

Every lesson in this journal begins with coloring pages. Your student may wish to color these pages while the lesson is read aloud. Most lessons also include a template with several empty boxes and writing lines. After each reading session, encourage your child to use the boxes and lines to record information he found interesting in the reading. Your child can create illustrations, diagrams, or short narrations of what he's learned. By doing this, your child's retention of the material will be increased significantly. Following this template for creative expression is another template for completing the notebooking assignment from the text. Colored pencils are encourages as they facilitate creativity and high quality work. Hands-on vocabulary activities are also provided for each lesson to help your student learn important astronomy terms.

Some experiments in the book require the student to use a Scientific Speculation Sheet. These sheets have been included in this notebooking journal. Drawings or pictures of the projects can be pasted onto the Scientific Speculation Sheets.

Scripture Copywork

Incorporating the Word of God in your science studies through Scripture Copywork will provide many benefits to your student. It will encourage stronger faith and memorization of Scripture, as well as better writing, spelling and grammar skills. The copywork is designed to be traced over and then recopied on the lines below the Scriptures.

Project Pages

If your student chooses to do one of the projects in the book or one of the Take It Further suggestions, he may wish to include a drawing or photograph on the Project Page provided for that lesson. This will remind him of his project and what he learned. It will also serve as a record of his learning.

Cut and Fold Miniature Books

At the back of this journal, your will find Cut and Fold Miniature Book craft activities for each lesson. These are entirely optional. These miniature books are designed to review the concepts learned in each lesson. Paste pages are included in this journal for each miniature book activity. The Paste Pages provide a place for your students to preserve and display their Cut and Fold Miniature Books. Instructions are included for cutting and assembling the miniature books.

Take It Further

The Take It Further suggestions are designed to give your student additional ideas and activities that might enhance his studies such as: experiments, hands-on activities, research and living book titles, as well as audio and video resources. Because these assignments are entirely optional, they are not included in the Suggested Schedule for completing the notebooking journal.

Field Trip Sheets

Your family may wish to further enhance your studies by visiting a science museum or perhaps a Planetarium. Field Trip Sheets are provided at the back of this notebooking journal to record your visits. you can make a pocket on the back of these sheets to hold any brochures or additional information your receive. Simply glue three edges (sides and bottom) of a half piece of construction paper to the bottom of the Field Trip Sheet.

Week	Day 1	Day 2
1	**Lesson 1 - What is Astronomy?** Read *T pp. 2-5* & Narrate Begin working on Coloring Pages about Astronomy *NJ pp. 10-11*	Read *T pp. 5-8* & Narrate Begin working on Fascinating Facts about Astronomy *NJ p. 12* Notebooking Activity: Mnemonic *T p. 9, NJ p. 13*
2	**Lesson 1 - What is Astronomy?** Scripture Copywork *NJ pp. 14-15*	Vocabulary Crossword *NJ p. 16* Project: Solar System *T pp. 9-10, NJ p. 17* What is Astronomy? Minibook *Appendix p. A 7*
3	**Lesson 2 - The Sun** Read *T pp. 12-17* & Narrate Begin working on Coloring Pages about the Sun *NJ pp. 20-21* Read *T pp. 18-19* Begin working on Fascinating Facts about the Sun *NJ pp. 22-23*	Mid-lesson Notebooking Activity: Why You Should Not Look at the Sun Speech *T p. 18, NJ p. 24* Mid-lesson Activity: Use a Magnifying Glass to Focus Heat *T p. 19*
4	**Lesson 2 - The Sun** Read *T pp. 20-25* & Narrate Notebooking Activity: Sun Collage *T p. 26, NJ p. 25* Scripture Copywork *NJ pp. 26-27*	Vocabulary Lift the Flap *NJ p. 29* Activity: Make a Solar Eclipse *T p. 26* Project: Pinhole Viewing Box *T pp. 27-28, NJ p. 32* Sun Minibooks *NJ Appendix pp. A 11, A 15*
5	**Lesson 3 - Mercury** Read *T pp. 30-34* & Narrate Begin working on Coloring Pages about Mercury *NJ pp. 36-37*	Read *T pp. 34-36* & Narrate Begin working on Fascinating Facts about Mercury *NJ pp. 38-39*
6	**Lesson 3 - Mercury** Project: Create Craters *T p. 37* Scripture Copywork *NJ pp. 40-41*	Vocabulary Puzzle Game *NJ p. 43* Project: Make a Model of Mercury *T p. 38, NJ p. 48* Mercury Minibook *NJ Appendix p. A 17*
7	**Lesson 4 - Venus** Read *T pp. 40-41* & Narrate Mid-lesson Activity: Make Some "Lava" *T p. 41* Begin working on Coloring Pages about Venus *NJ pp. 51-52*	Read *T pp. 42-47* & Narrate Begin working on Fascinating Facts about Venus *NJ pp. 53-54*
8	**Lesson 4 - Venus** Notebooking Activity: Comic Strip *T p. 48, NJ p. 55* Scripture Copywork *NJ pp. 56-57*	Venus Vocabulary Story *NJ p. 58* Project: Learn How Radar is Used *T p. 49-51, NJ p.59* Venus Minibook *NJ Appendix p. A 19*
9	**Lesson 5 - Earth** Read *T pp. 52-57* & Narrate Begin working on Coloring Pages about Earth *NJ pp. 65-66*	Read *T pp. 57-63* & Narrate Begin working on Fascinating Facts about Earth *NJ pp. 67-68*
10	**Lesson 5 - Earth** Notebooking Activity: Advertisement *T p. 63, NJ p. 69* Scripture Copywork *NJ pp. 70-71*	Vocabulary Lift the Flap *NJ p. 73* Project: Make a Compass *T p. 64, NJ p. 76* Earth Minibooks *NJ Appendix pp. A 25, A 27*
11	**Lesson 6 - Moon** Read *T pp. 66-70* & Narrate Begin working on Coloring Pages about the Moon *NJ pp. 79-80*	Read *T pp. 70-74* & Narrate Begin working on Fascinating Facts about the Moon *NJ pp. 81-82*
12	**Lesson 6 - Moon** Notebooking Activity: Chart the Moon *T p. 74, NJ p. 83* Scripture Copywork *NJ pp. 84-85*	Moon Phases *NJ p. 86* Project: Make a Telescope *T pp. 75-76, NJ p. 87* Moon Minibook *NJ Appendix p. A 29*
13	**Lesson 7 - Mars** Read *T pp. 78-82* & Narrate Begin working on Coloring Pages about Mars *NJ pp. 91-92*	Read *T pp. 82- 85* & Narrate Begin working on Fascinating Facts about Mars *NJ pp. 93-94*
14	**Lesson 7 - Mars** Notebooking Activity: Going to Mars *T p. 85, NJ p. 95* Scripture Copywork *NJ pp. 96-97*	Vocabulary Crossword *NJ p. 98* Project: Design a Mars Community *T p. 86, NJ p. 99* Project: Build the Olympus Mons Volcano *T pp. 87-88, NJ p. 99* Mars Minibooks *NJ Appendix p. A 33*
15	**Lesson 8 - Space Rocks** Read *T pp. 90-94* & Narrate Begin working on Coloring Pages about Space Rocks *NJ pp. 103-104*	Read *T pp. 95-99* & Narrate Begin working on Fascinating Facts about Space Rocks *NJ p. 105*

*Page numbers for the astronomy text are indicated by *T p.* Page numbers for the notebooking journal are indicated by *NJ p.*

Week	Day 1	Day 2
16	**Lesson 8 - Space Rocks** Notebooking Activity: Comets, Meteors, Asteroids *T p. 100, NJ p. 106* Notebooking Activity: Meteor Showers *T p. 100, NJ p. 107* Scripture Copywork *NJ pp. 108-109*	Space Rocks Vocabulary Story *NJ p. 110* Project: Create a Scale Model Solar System *T pp. 101-102, NJ p. 111* Space Rocks Minibook *NJ Appendix p. A 35*
17	**Lesson 9 - Jupiter** Read *T pp. 104-106* & Narrate Begin working on Coloring Pages about Jupiter *NJ pp. 114-115*	Read *T pp. 107-110* & Narrate Begin working on Fascinating Facts about Jupiter *NJ p. 116*
18	**Lesson 9 - Jupiter** Notebooking Activity: Make a Jovian Newspaper *T p. 110-111, NJ p. 117* Scripture Copywork *NJ pp. 118-119*	Jupiter Vocabulary *NJ p. 120* Project: Make a Hurricane Tube *T p. 112, NJ p. 121* Jupiter Minibook *NJ Appendix p. A 39*
19	**Lesson 10 - Saturn** Read *T pp. 114-116* & Narrate Begin working on Coloring Pages about Saturn *NJ pp. 124-125*	Begin working on Fascinating Facts about Saturn *NJ pp. 126-127* Read *T pp. 116-118* & Narrate
20	**Lesson 10 - Saturn** Notebooking Activity: Make a Venn Diagram *T p. 118, NJ p. 128* Vocabulary Crossword *NJ p. 129*	Scripture Copywork *NJ pp. 130-131* Project: Make a Centaur Rocket *T pp. 119-120, NJ p. 132* Saturn Minibook *NJ Appendix p. A 41*
21	**Lesson 11 - Uranus & Neptune** Read *T pp. 122-124* & Narrate Begin working on Coloring Pages about Uranus & Neptune *NJ pp. 135-136*	Read *T pp. 125-127* & Narrate Begin working on Fascinating Facts about Uranus & Neptune *NJ p. 137*
22	**Lesson 11 - Uranus & Neptune** Notebooking Activity: Create a Play about the Discovery of Uranus *T p. 128-129, NJ p. 138* Neptune & Uranus Match Up *NJ p. 139*	Scripture Copywork *NJ pp. 142-143* Project: Make Clouds *T p. 130, NJ p. 144* Uranus & Neptune Minibooks *NJ Appendix p. A 43*
23	**Lesson 12 - Pluto & Kuiper Belt** Read *T pp. 132-135* & Narrate Begin working on Coloring Pages about Pluto & Kuiper Belt *NJ pp. 147-148*	Read *T pp. 136-138* & Narrate Begin working on Fascinating Facts about Pluto & Kuiper Belt *NJ p. 149*
24	**Lesson 12 - Pluto & Kuiper Belt** Notebooking Activity: Pluto Debate *T p. 139, NJ p. 150* Notebooking Activity: Pluto Debate Minibook *T p. 139, NJ Appendix p. A 47* Vocabulary Puzzle Game *NJ p. 151*	Scripture Copywork *NJ pp. 156-157* Project: Make Ice Cream! *T p. 140, NJ p. 158* Pluto & Kuiper Belt Minibooks *NJ Appendix pp. A 47, A 48*
25	**Lesson 13 - Stars & Galaxies** Read *T pp. 142-148* & Narrate Begin working on Coloring Pages about Stars & Galaxies *NJ pp. 161-162*	Begin working on Fascinating Facts about Stars & Galaxies *NJ pp. 163-164* Read *T pp. 149-155* & Narrate Notebooking Activity: Research & Illustrate Constellations *T p. 156, NJ p. 165*
26	**Lesson 13 - Stars & Galaxies** Notebooking Activity: Locate Constellations *T p. 156, NJ p. 166* Stars & Galaxies Identification *NJ p. 167*	Scripture Copywork *NJ pp. 168-169* Project: Make an Astrometer *T p. 157, NJ p. 170* Project: Create a Constellation Planetarium *T p. 158, NJ p. 170* Stars & Galaxies Minibook *NJ Appendix p. A 49*
27	**Lesson 14 - Space Travel** Read *T pp. 160-167* & Narrate Begin working on Coloring Pages about Space Travel *NJ pp. 173-174*	Begin working on Fascinating Facts about Space Travel *NJ p. 175* Read *T pp. 167-169* & Narrate
28	**Lesson 14 - Space Travel** Notebooking Activity: Let's Visit the Planets! *T p. 170-171, NJ p. 176* Vocabulary Puzzle Game *NJ p. 177*	Scripture Copywork *NJ pp. 182-183* Project: Build a Model Space Station *T p. 172, NJ p. 184* Space Travel Minibook *NJ Appendix p. A 53* Solar System Review Minibook *NJ Appendix p. A 59*

*Page numbers for the astronomy text are indicated by *T p.* Page numbers for the notebooking journal are indicated by *NJ p.*

This journal belongs to:
_____

After they had heard the king, they went on their way, and the star they had seen in the east went ahead of them until it stopped over the place where the child was.   Matthew 2:9

By the word of the LORD the heavens were made, their starry host by the breath of his mouth.

Psalm 33:6

Fascinating Facts
About
Astronomy
Lesson 1

Mnemonic
Lesson 1

Mnemonic with Pluto

Mercury	Venus	Earth	Mars	Jupiter	Saturn	Uranus	Neptune	Pluto

Write Your Sentence Here:

_____

- -

_____

Mnemonic without Pluto

Mercury	Venus	Earth	Mars	Jupiter	Saturn	Uranus	Neptune

Write Your Sentence Here:

_____

- -

_____

The heavens are telling of the glory of God; and their expanse is declaring the work of His hands.

Psalm 19:1

The heavens are telling of the glory of God; and their expanse is declaring the work of His hands.

Psalm 19:1

Vocabulary Crossword
Lesson 1

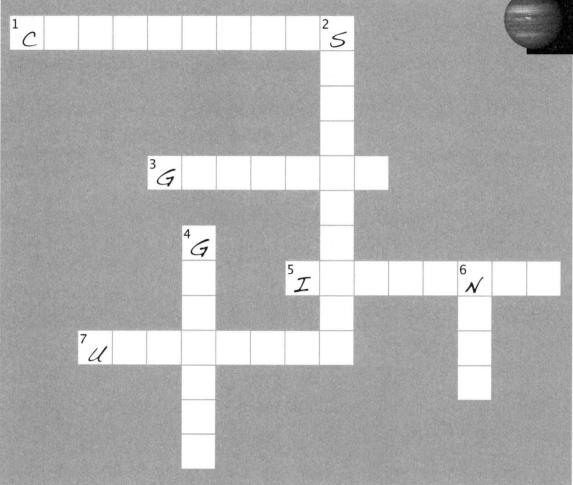

NASA

GALILEO

COPERNICUS

INSTINCT

UNIVERSE

STONEHENGE

GRAVITY

Across

1. A scientist who discovered that the earth revolves around the sun, rather than the sun revolving around the earth, as was believed at the time.
3. A scientist who built telescopes and studied astronomy.
5. A special gift God gives to creatures, causing them to behave in a certain manner that is helpful to their survival, such as with birds flying south for the winter.
7. Everything that exists in space, including the earth, planets, sun, and stars.

Down

2. An ancient monument in England that may have been used to predict the arrival of spring and other seasons.
4. A physical force causing objects to pull on other objects, such as with the sun pulling on the earth.
6. America's space agency, called the National Aeronautics and Space Administration.

My Astronomy Project
Lesson 1

What I did:

What I learned:

Astronomy Minibooks
Lesson 1

Paste your What is Astronomy?
Matchbook onto this page.

TAKE IT FURTHER
LESSON 1

Hanging Solar System in a Box

You will need:
Clay
Thread
A copy paper box (found at office supply stores)
A sharp pencil
Duct tape (or other strong tape)

Using the sizes for the planets on page 10 of *Exploring Creation with Astronomy*, you can create a smaller scale solar system using clay. Simply reduce the sizes to millimeters or centimeters and model each planet out of clay, forming balls. If you are not worried about making it perfectly to scale, you can make the smaller planets a little larger for ease of handling. You will want to insert a piece of thread in the center of each planet before it dries, using a threaded needle. Hang the planets inside the box by using the sharp pencil to make holes in the top of the box, inserting the free ends of the thread through the holes and taping them down with duct tape.

Solar System Mobile

You will need:
Cardstock
Colored markers
Tacks
String
Tape

An even simpler idea is to create a solar system mobile using cardstock and markers. Reduce the sizes on page 10 of *Exploring Creation with Astronomy* to centimeters and draw circles on the cardstock to represent each planet. For larger planets, you will need to piece two pages together. Color the planets and cut them out (or use colored construction paper). Tape the string to them and tack the paper planets to the ceiling as you would the balloons.

Make a Model of Stonehenge

Make a model of Stonehenge with clay. Look at pictures on the Internet to see how it is formed from different perspectives. Be sure to photograph it for your Astronomy Project Page!

Book Suggestions

Astronomy (DK Eyewitness) by Kristen Lippencott. This comprehensive resource details the history of astronomy, important astronomers and what is known about the planets, stars and sun. (ages 9-12)
Along Came Galileo by Jeanne Bendick. Here's Galileo's inspiring story told with whimsical illustrations and engaging text.
Nicolaus Copernicus: The Earth Is a Planet by Dennis Fradin. This picture book biography includes many interesting facts about this fascinating scientist.
Signs and Seasons by Jay Ryan. An introduction to classical astronomy with a biblical basis.
Star Seeker: A Journey to Outer Space by Theresa Heine. This colorful and lyrical picture book describes an imaginary trip to outer space where planets, constellations and other heavenly bodies are discovered. (ages 6-9)

"Then THE RIGHTEOUS WILL SHINE FORTH AS THE SUN in the kingdom of their Father. He who has ears, let him hear.

Matthew 13:43

And night will be no more. They will need no light of lamp or sun, for the Lord God will be their light, and they will reign forever and ever.   Revelation 22:5

Fascinating Facts

About The Sun

Lesson 2

Fascinating Facts about The Sun
Lesson 2

Why You Should not Look at the SUN

SUN COLLAGE

From the rising of the sun unto the going down of the same the LORD's name is to be praised.

Psalm 113:3

From the rising of the sun unto the going down of the same the LORD's name is to be praised.

Psalm 113:3

VOCABULARY LIFT THE FLAP
LESSON 2

Tear out this page. Cut out each word and match it to the correct definition on the following page. Then, place glue along the top edge of the back of each word and glue above the line on each definition. Once the glue is dry, fold back the word to reveal the definition.

ORBIT	ECLIPSE
THERMONUCLEAR FUSION	ATMOSPHERE
SOLAR FLARES	SUNSPOTS
ROTATE	AURORAS

SUN VOCABULARY

Glue correct word above this line, then fold back.

When the moon is between the earth and the sun, completely covering the sun, causing the moon to cast a shadow upon the earth.

Glue correct word above this line, then fold back.

The path that an object or planet follows as it moves around the sun.

Glue correct word above this line, then fold back.

The layers of gases and mist that cover a planet, such as the earth.

Glue correct word above this line, then fold back.

Nuclear reactions (controlled explosions) that happen on the sun that give it energy and power.

Glue correct word above this line, then fold back.

Large fires that burst out millions of miles from the sun, releasing energy into the solar system.

Glue correct word above this line, then fold back.

Dark patches on the sun which are cooler than the rest of the sun.

Glue correct word above this line, then fold back.

Natural, colorful light displays in the sky that can be caused by solar flares.

Glue correct word above this line, then fold back.

To turn around a fixed point, the way a top spins in place or a planet spins on its axis.

My Sun Projects
Lesson 2

What I did:

What I did:

What I learned:

What I learned:

Sun Minibooks
Lesson 2

Paste your Sun Wheel and Sun Minibooks onto this page.

TAKE IT FURTHER
Lesson 2

Learn about Shadows and the Sun

You will need:
A dowel
Your yard
A sunny day
Measuring tape

You can learn about the way the shadows change throughout the day. On a sunny day, in the early morning, place a dowel in the ground and measure the length of the shadow, noting the time and where the shadow falls in relation to the sun and the earth. Every few hours, go outside and measure the shadow, noting the time. You can create a simple chart with the time and length of the shadow. This will be a tool you can use to record your observations.

Purchase Solar Glasses

You can purchase solar glasses with which to view the sun directly. One can see sun spots and solar flares through these glasses. They are the only safe glasses to use when looking at the sun. Never look at the sun without special devices made for this purpose. To find them on the Internet, use the search term, "eclipse glasses."

Learn about Color

You will need:
A coffee filter
A water-based black marker
A cup of rubbing alcohol

This experiment will help you see that black is all the colors of the rainbow mixed together. Fold a coffee filter in half, creating a half circle. Draw a thick line in an arch just below the ruffled edge, creating a black rainbow shape on your coffee filter. Fold the coffee filter in half again, making a pizza slice shape. Put the pointed end in a cup of alcohol, keeping the black arch above the alcohol line. As the alcohol absorbs into the filter, it will spread up to the black arch and begin to separate the colors.

Solar Prints

Place various objects (scissors, rubber bands, pencils) on a piece of blue and a piece of red construction paper, and place the papers in a sunny spot outside. At the end of the day, remove the objects. What caused the paper to fade around the objects? Which paper faded more? Why do you think that is, based on what you learned about color and the sun?

TAKE IT FURTHER
LESSON 2

Make a UV Detecter

Just where are the sun's UV rays? You know, those rays that can cause a sunburn or damage your eyes? You cannot see them, but with the help of a UV detector bracelet, you can learn more about them. Let's give it a try!

You will need:

UV sensitive beads (12-15 will make a bracelet, and you can buy them in bulk online.)
A piece of string or lacing (long enough to make a bracelet that slips off easily)
A flashlight
A sunny day

String as many beads as you'd like onto the lacing and tie a knot to secure it. Shine an ordinary flashlight onto the beads and record what happens. Now, go outside and record what happens to the beads as you place them in various spots: in the direct sun, under a shady tree, in the shade of a building. Does the response change as you walk from a sunny spot into the darkest part of the shade? Use the beads to set up scientific experiments to answer the following questions:

Is there really a difference in sunscreens with different SPF's? You can apply sunscreen directly to the beads or on a sheet of acetate that you hold over the beads. If you have a lot of beads, you can put them in zippered plastic bags and apply the different sunscreens to the bags. Be sure to try other opaque substances, like shortening.

Do all sunglasses protect your eyes?

Does the intensity of UV rays change throughout the day?

Book and DVD Suggestions

Sun by Steve Tomecek. This brightly colored picture book provides a gentle introduction to the science of the sun. (ages 4-8)
The Sun: Our Nearest Star by Franklyn Branley. This colorful picture book explores the sun and the energy it produces, allowing plants and animals to exist on earth. (ages 4-8)
Sun up, Sun Down: The Story of Day and Night by Jacqui Bailey. This picture book follows the sun from dawn to dusk and explains how light rays travel, how shadows are formed, and how the moon lights up the night sky. (ages 6-9)
The Sun by Elaine Landau. The lively narrative and liberal use of photographs in this book will capture readers while introducing them to the power of the sun. (ages 9-12)
A Look at the Sun by Ray Spangenburg and Kit Moser. This book contains engaging text and photographs that explore in depth the history of man's ideas about and observations of the sun. (ages 10+)
Bill Nye the Science Guy: the Sun; distributed by Disney Educational Productions 2008. Zaney Bill Nye explores solar flares, eclipses, sunspots, fusion and solar energy. He takes viewers on a field trip to a solar energy farm outside of Sacramento, California. (all ages)

"He gives wisdom to the wise and knowledge to the discerning. He reveals deep and hidden things; he knows what lies in darkness, and light dwells with him.
Daniel 2:21-22

The LORD will guide you always; he will satisfy your needs in a sun-scorched land and will strengthen your frame. You will be like a well-watered garden, like a spring whose waters never fail.

Isaiah 58:11

Fascinating Facts
About
Mercury
Lesson 3

Mariner 10

How old was your mom when Mariner 10 first visited Mercury? If she was born after 1974, try to find out how old your grandma was.

Fascinating Facts

About Mercury
Lesson 3

"Every good gift and every perfect gift is from above, and comes down from the Father of lights."

James 1:17

"Every good gift and every perfect gift is from above, and comes down from the Father of lights."

James 1:17

Vocabulary Puzzle Game

TRANSIT

GASEOUS

ASTEROIDS

Planets that are "earth-like," having a solid surface upon which you can stand.

A spacecraft that travels without a person inside.

A dent on the surface of a planet or moon caused by the impact of an asteroid.

Cut out each puzzle piece on this page and the next. Match each vocabulary word puzzle piece to the correct definition puzzle piece. Cut out this rectangle and glue it to your Puzzle Page along the bottom and side edges to create a pocket. After you've played your Vocabulary Puzzle Game a few times, place all your puzzle pieces in the pocket for safe keeping.

-Glue along this edge-

-Glue along this edge-

-Glue along this edge-

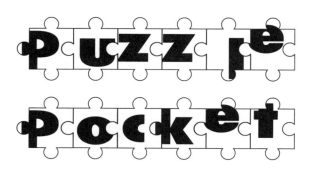

TERRESTRIAL

CRATER

Rocks that orbit in space, sometimes crashing into planets and other satellites.

Planets that are not solid, but instead are made of gas.

This word means to pass over. We use this word to describe when Mercury passes between the earth and the sun.

TRANSIT

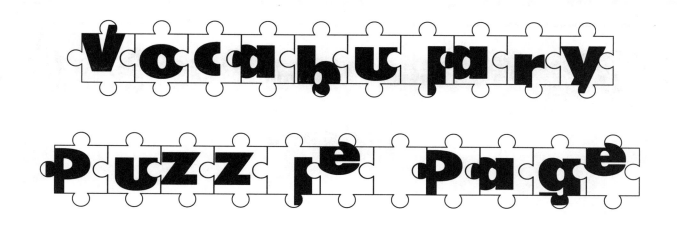

Paste your puzzle pocket here.

My Mercury Project
Lesson 3

What I did:

What I learned:

Mercury Minibooks
Lesson 3

Paste your Mercury
Minibooks onto this page.

TAKE IT FURTHER
LESSON 3

ALTERNATIVE IMPACT CRATER EXPERIMENT

In this experiment, you will take the book's crater activity even further. You will study how the impact crater's size is affected by the size of the object and the height from which it fell. You will only need 3 pebbles. You can also use balls, marbles, or any other small object – it does not have to be round.

In addition to the experiment items listed in the book (p. 37), you will need:

Dry powder paint
A flour sifter or sieve
A balance or postal scale for weighing the pebbles
A ruler
Tweezers or small kitchen tongs
An old newspaper to protect the work area

Put down some old newspaper or a table covering to catch any stray powder from the bowl. On top of the flour, use the flour sifter to sift a thin layer of colored powder paint. Now, find the weight and diameter of each of your pebbles (you can use other items like different sized balls or marbles if pebbles are not available). Record these numbers on a chart.

Take the smallest pebble, hold it one foot above the bowl and drop it into the bowl. Carefully, with tweezers or tongs, remove the pebble and measure the depth and width, as well as the average length of the rays that project out from the impact. The rays are white streaks of flour radiating from the crater. If it's easier for you, you can measure the depth by measuring the pebble and then measuring how far it protrudes from the surface. Subtracting that number from the pebble will give you the depth of the impact. Record all your data on the table below. Write down anything else you notice about the impact.

Now repeat the procedure with the same pebble, dropping it first two feet, then four feet and then, standing on a chair, from six feet above the bowl. Create a table like the one below for each height and record your measurements on the correct table after each impact.

Repeat the same procedure with all the pebbles, from biggest to smallest. When you are done, write down the total (adding all the numbers together) and then find the average by dividing the total by three. After you have studied your tables, make a conclusion about how big of an average impact you might have if an object fell from three feet, five feet and six feet. Test your hypothesis!

Record your data on the table below:

1 Foot	1st Object	2nd Object	3rd Object	Total	Average
Crater Width					
Crater Depth					
Ray Length					
Observations					

The path of the righteous is like the first gleam of dawn, shining ever brighter till the full light of day.   Proverbs 4:18

The highest heavens belong to the LORD, but the earth he has given to man.
Psalm 115:16

Fascinating Facts

About Venus

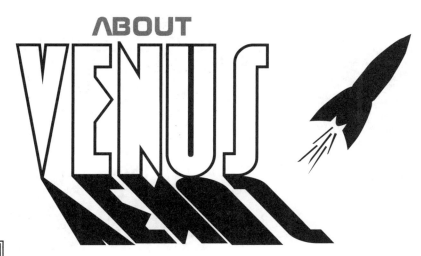

Lesson 4

Fascinating Facts

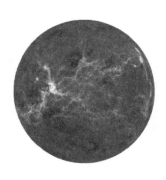

Lesson 4

VENUS COMIC STORY

And the light shines in darkness; and the darkness comprehended it not.

John 1:5

And the light shines in darkness;
and the darkness comprehended it
not.

John 1:5

Venus Vocabulary Story
Lesson 4

I looked up in the sky right before the sun went down and saw a little star brightly shining in the sky near the horizon. I pointed to it and said, "Look! The first star of the evening." My mom said, "That's not a star. That's the planet _____ . But many people call it the _____ _____ when they see it in the evening. It is very close to the sun, so we see it when the sun is coming up or going down.

My mom explained that Venus goes through phases, just like the moon. So, when we look at Venus through a telescope, it might be perfectly round, like a full moon. Or it might be in the shape of a _____ , like when we see only a sliver of the moon. The planet doesn't actually change shapes, it's just the way the sun's light hits it.

A few weeks later, I woke up at the crack of dawn and looked out my window. Guess what I saw! I saw Venus again! It was shining so early in the morning. I remembered my mom told me that when Venus is seen early in the morning, we call it the _____ _____ .

I wish I could travel in a spaceship to Venus and see what it looks like up close. My mom says that would be very dangerous because the surface of Venus has many volcanoes on it. The volcanoes spill out _____ , which is hot, liquid rock. When the liquid dries, it leaves really jagged rocks all over the place. So, Venus isn't as pretty as the earth, but I still think it would be exciting to see it in real life!

Choose from the words below to fill in the blanks in the Venus Vocabulary Story.

Lava   Morning Star   Crescent
Venus   Evening Star

LESSON 4

What I did:

What I learned:

RADAR CHART FOR VENUS EXPERIMENT

1	2	3	4	5	6	7	8	9	10
11	12	13	14	15	16	17	18	19	20
21	22	23	24	25	26	27	28	29	30
31	32	33	34	35	36	37	38	39	40
41	42	43	44	45	46	47	48	49	50
51	52	53	54	55	56	57	58	59	60
61	62	63	64	65	66	67	68	69	70
71	72	73	74	75	76	77	78	79	80

RADAR CHART FOR VENUS EXPERIMENT

1	2	3	4	5	6	7	8	9	10
11	12	13	14	15	16	17	18	19	20
21	22	23	24	25	26	27	28	29	30
31	32	33	34	35	36	37	38	39	40
41	42	43	44	45	46	47	48	49	50
51	52	53	54	55	56	57	58	59	60
61	62	63	64	65	66	67	68	69	70
71	72	73	74	75	76	77	78	79	80

LESSON 4

Paste your Venus Volcano onto
this page.

TAKE IT FURTHER
LESSON 4

Alternate Venus Radar Ideas

Instead of using Plaster of Paris, you can build your Venus terrain using any of the following items:

Legos
Blocks
Toys and stuffed animals
Newspaper and duct tape

You will want to make certain that the objects you choose don't move around in the box. You might want to tape them to the bottom of the box.

Book Suggestions

Brightest in the Sky: the Planet Venus by Nancy Loewen. This picture book explores earth's closest and brightest neighbor, the planet Venus. (ages 6-9)

Venus by Elaine Landau. Using lively narrative and photographs, this book describes the characteristics of Venus and the history of its space missions. (ages 9-12)

A Look at Venus by Ray Spangenburg and Kit Moser. Using engaging text and photographs, this book describes the earliest ideas about and observations of moonless Venus. (ages 10+)

For this is what the LORD says--he who created the heavens, he is God; he who fashioned and made the earth, he founded it; he did not create it to be empty, but formed it to be inhabited--he says: "I am the LORD, and there is no other."
Isaiah 45:18

from his dwelling place he watches
all who live on earth—
he who forms the hearts of all,
who considers everything they do.
Psalm 33:14-15

Fascinating Facts

About

Earth
Lesson 5

COPYWORK

For by Him all things were created, both in the heavens and on earth...all things have been created by Him and for Him.
Colossians 1:16

COPYWORK

For by Him all things were created, both in the heavens and on earth...all things have been created by Him and for Him.

Colossians 1:16

Vocabulary Lift the Flap
Lesson 5

Tear out this page. Cut out each earth and match its title to the correct definition on the following page. Then, place glue along the top edge of the back of each earth and glue above the line on each definition. Once the glue is dry, fold back the earth to reveal the definition.

Earth Lift the Flap
Lesson 5

Glue correct word above this line, then fold back
___

The solid outermost layer of the earth.

Glue correct word above this line, then fold back
___

A large magnetic field that surrounds the earth.

Glue correct word above this line, then fold back
___

The imaginary line or circle around the center of the earth.

Glue correct word above this line, then fold back
___

The solid center of the earth.

Glue correct word above this line, then fold back
___

The layer of the earth just below the earth's crust.

Glue correct word above this line, then fold back
___

The only planet in our solar system that can support life.

My Earth Project
LESSON 5

What I did:

What I learned:

LESSON 5

Paste your Earth Layers Book and Earth Lift Book onto this page.

TAKE IT FURTHER
Lesson 5

Edible Earth

Can you remember the layers of the earth? Here's a fun and tasty way to help you! **You will need:**

Rice Krispy Treats (the recipe is on the back of a Rice Krispies box)
A red gumdrop
Chocolate chips (melted)

Begin creating your edible earth with the red gumdrop. It represents earth's inner layer, the core. Next, wrap a layer of Rice Krispy Treats around the gumdrop. This represents earth's mantle. Last, cover your creation with a layer of chocolate. This represents earth's outer layer, the crust. Try to shape your edible earth into a smooth ball. As you bite into your earth, see if you can name the layers!

Book Suggestions

You're Aboard Spaceship Earth by Particia Lauber. Using simple text and colorful illustrations, this picture book explores the life-sustaining natural resources of planet earth.

Here in Space by David Milgrim. Following the space explorations of a young boy, readers will be challenged to look at earth from a new perspective. (ages 4-8)

On Earth by Brian Karas. This beautifully illustrated picture book explores earth's cycles of rotation and revolution, and how they affect our lives through the days and seasons. (ages 4-8)

What's so Special About Planet Earth? by Robert E. Well. Using simple text and colorful illustrations, this picture book explores the seven other planets in our solar system, explaining why earth is the only one that can sustain human life. (ages 4-8)

The Librarian Who Measured the Earth by Kathryn Lasky. This picture book covers the life of Eratosthenes of Cyrene, a geographer who estimated the circumference of the earth. (ages 6-10)

Magic School Bus Inside the Earth by Joanna Cole. Containing uncluttered text and illustrations, this book is full of exciting adventures from the school yard to a volcano and back. (ages 6-10)

Journey to the Center of the Earth by Jules Verne. This chapter book takes young readers on an incredible journey, from Iceland's frozen tundra down into underground prehistoric worlds and back up again through the fires of an erupting volcano. (ages 9-12)

Our Home Planet: Earth by Nancy Loewen. With informative text and bright illustrations, this picture book explores our home planet, earth. (ages 6-9)

Earth by Elaine Landau. The lively narrative and liberal use of photographs in this book will capture readers while introducing them to our very own home, planet earth. (ages 9-12)

A Look at Earth by John Tabak. This book's engaging text and photographs describe in depth earth's composition, environment and biomes. (ages 10+)

The Earth is Painted Green: a Garden of Poems About Our Planet; edited by Barbara Brenner. This playfully illustrated picture book presents a collection of poems from around the world about life on earth. (all ages)

"You are the light of the world. A city on a hill cannot be hidden."
Matthew 5:14

This is what the LORD says, he who appoints the sun to shine by day, who decrees the moon and stars to shine by night, who stirs up the sea so that its waves roar-the LORD Almighty is his name. Jeremiah 31:35

Fascinating Facts About The Moon
Lesson 6

Fascinating Facts
About
The Moon
Lesson 6

Chart The Moon

He made the moon for the seasons; the sun knows the place of its setting.
Psalm 104:19

He made the moon for the seasons; the sun knows the place of its setting.
Psalm 104:19

Moon Phases
Lesson 6

_____
- - - - - - - - - - - - - - -
_____

_____
- - - - - - - - - - - - - - -
_____

_____
- - - - - - - - - - - - - - -
_____

_____
- - - - - - - - - - - - - - -
_____

Choose from the words below to label the moons.

Gibbous   Quarter
Crescent   Full

My Moon Project
Lesson 6

What I did:

What I learned:

Moon Minibook
Lesson 6

Paste your Moon Layered
Book onto this page.

88

TAKE IT FURTHER
Lesson 6

Draw the Moon

Go outside with a pair of binoculars. To still the binoculars, set them on a tripod or steady them on the back of a solid surface, such as a chair. Center the binoculars on the moon and draw the moon with crayons as you see it through the binoculars. You can glue your drawing onto the Moon Project Page.

Make Moon Cookies

To further the activity above, you can bake large round cookies and attempt to recreate the moon's surface with icing.

Earth and Moon to Scale Activity

* This activity is a guessing game designed for your parent to do with you. Have your parent follow the instructions below.

You will need:
Different sized balls - One should be 1/4th the size of the others (read the full instructions before gathering the balls).

Have your children guess which two balls accurately represent the size of the earth to the moon. After they have made their guess, tell them that four moons laid side by side equal the diameter of the earth (be sure you have two balls that can accurately represent the earth and the moon). Now, have them search for the two balls that would be a scale model of the earth and moon. Next, with the two balls, have your students guess how far away the moon is from the earth. Separate the two balls to that distance, reminding your students what it means to be "to scale" and how the size is reduced based on the size of the balls. The actual distance between the earth and moon would be close to putting 30 earths side by side, or 110 moons side by side. Show your students the actual size and distance difference between the earth and moon.

Moon Jumps

On the moon, because the gravity is weaker than the earth's gravity, you could jump six times as high as you can on the earth. Let's find out exactly how high you could jump on the moon! You will need masking tape and a yard stick or measuring tape. With a piece of masking tape in your hand, jump up as high as you can next to a wall and stick the masking tape on the wall at that height. Measure how high you were able to jump (or how high you were able to reach in your jump). Then take that number and multiply it by six. That's how high you could jump on the moon!

Book and DVD Suggestions

Moon (DK Eyewitness) by Jacqueline Mitton. This is a comprehensive guide to the moon, examining in depth its physical characteristics and the history of its space missions. (ages 9-12)
The Moon and You by Edwin C. Krupp. This book's engaging narrative and pencil-drawing illustrations capture the reader while describing the moon: its phases, rotation, effect on earth's tides and more.
Bill Nye the Science Guy: the Planets by Disney Educational Productions 2003. Zany Bill Nye explores the physical characteristics and phases of earth's natural satellite, the moon. (DVD, all ages)

TAKE IT FURTHER
LESSON 6

Lunar Landing Module

Did you know that manned spacecrafts are designed to land softly – whether on the moon or when they return back to earth. They are designed that way in order to protect the astronauts inside and also the spacecraft. Can you design such a device? Let's try! If you are successful, you will be able to drop your spacecraft on the floor without harming your astronauts or the module.

You will need:
An 8 oz. paper cup (for the cockpit)
Mini marshmallows (for astronauts and possible shock absorbers)
A 5" x 5" cardboard square (for the platform)
Index cards (fan folded into a spring)
Tape
Rubber Bands
Scissors
Anything that would absorb the shock of impact: foam, cotton balls, marshmallows, straws (which can be assembled into a flexible design), index cards folded like a fan, etc.

Tape the bottom of your astronauts' cockpit (cup) to the cardboard platform. The only design restriction is that your astronauts (two mini marshmallows) cannot in any way be constrained inside the landing module. They need to be able to fall out if the module tips over. How you choose to design the bottom shock absorbing part is up to you. Remember that scientists and inventors rarely get their inventions right the first time they try. A successful design is often the result of many trials. One way to do it is pictured below. Try it and then see if you can come up with a better design. You will want to closely watch what happens when your module lands and then modify it to change the outcome. If the module tips over, then you might try moving your cockpit more to the center of the platform. Or perhaps add more shock absorbers to that side.

Start off dropping your module from a height of 2 feet. If that's successful, try 3 feet. Try this activity with your friends, and see who can come up with the design that can be dropped the farthest with the most success!

Images and Experiment Courtesy of Cindy Frick

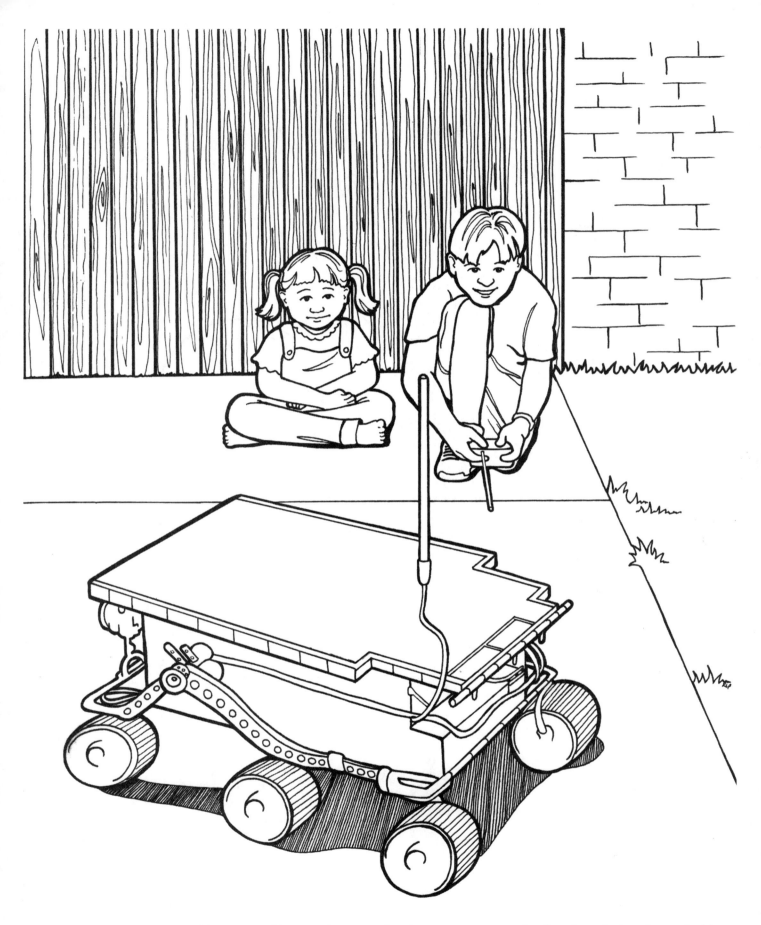

"It is I who made the earth, and created man upon it. I stretched out the heavens with My hands, and I ordained all their host."

Isaiah 45:12

And when you look up to the sky and see the sun, the moon and the stars--all the heavenly array--do not be enticed into bowing down to them and worshiping things the LORD your God has apportioned to all the nations under heaven.

Deuteronomy 4:19

Fascinating Facts
About
Mars
Lesson 7

Fascinating Facts

About
Mars
Lesson 7

GOING TO MARS
LESSON 7

COPYWORK

Oh LORD, our Lord, how majestic is Your name in all the earth, Who have displayed Your splendor above the heavens!

Psalm 8:1

COPYWORK

Oh LORD, our Lord, how majestic is Your name in all the earth, Who have displayed Your splendor above the heavens!

Psalm 8:1

Vocabulary Crossword
Lesson 7

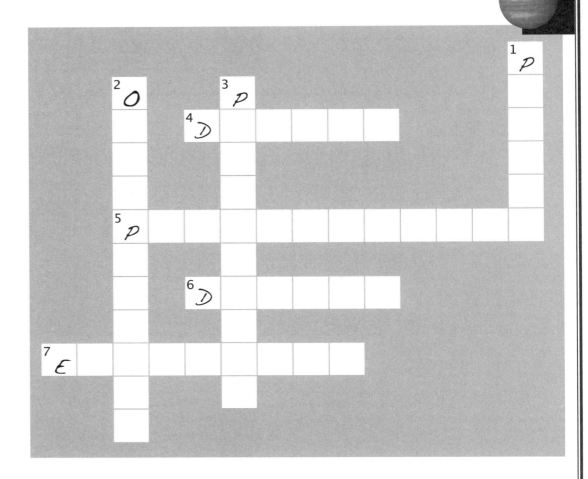

OLYMPUS MONS
ECOSYSTEM
PHOBOS
POLAR ICECAPS
DRY ICE
PERMAFROST
DEIMOS

Across

4. The smaller moon orbiting Mars.
5. Large slabs of ice at the North and South Poles of the planet Mars.
6. Frozen carbon dioxide.
7. A special habitat that can sustain life. On Mars, an artificial _____ would be an enclosed structure sealing in oxygen, allowing the sun's light to penetrate, containing all that is needed to sustain life on Mars.

Down

1. The larger moon orbiting Mars.
2. The largest volcano in our solar system.
3. Ground water that is considered permanently frozen.

MY MARS PROJECTS
LESSON 7

What I did:

What I did:

What I learned:

What I learned:

MARS MINIBOOKS
LESSON 7

Paste your Mars Wheels onto this page.

100

TAKE IT FURTHER
Lesson 7

Why is the Red Planet Red? Experiment

Scientists think Mars was once flowing with water and lava bursting from active volcanoes. They believe the red color of the planet was produced by rusting iron. This iron could only rust in a wet environment. In this activity, you will come to understand how a planet that had a great deal of iron could become red.

You will need:
Two pie pans
Light-colored sand
Steel wool
Scissors
A magnifying glass
Water

Fill both pie pans with sand. Cut the steel wool into tiny pieces and mix it with the sand. Pour water into one pie pan until it is very wet, but not flooded. Keep the other pie pan completely dry. The wet pie pan is your variable pan. The dry pie pan is your constant pan. Study both pans with the magnifying glass. Make a guess about what will happen at the end of the week. Make certain the variable pie pan remains damp all week. Each day, check on the pie pans and observe the differences. Study with a magnifying glass at the end of the week to observe both pans. If you'd like, you can continue this experiment until the end of the month.

Make an Edible Mars Rover

Using pictures found on the Internet, recreate a Mars Rover (such as Sojourner or Pathfinder) using tasty treats.

Sojourner

You will need:
Icing to put the rover together
Graham crackers
Round cookies for the wheels
Twizzlers and other assorted candy to make different components

Take a picture before you eat it! Be sure to glue your picture onto the Mars Project Page.

Book Suggestions

Messages From Mars by Loreen Leedy. This fun picture book follows the adventures of six children who travel through space to Mars, reporting their findings back to earth by email. (ages 6-9)
Cars on Mars: Roving the Red Planet by Alexandra Siv. This book is filled with detailed photographs and follows the course of NASA's Mars Exploration Rover Mission of 2003. (ages 9-12)
The Adventures of Sojourner: the Mission to Mars that Thrilled the World by Susi Trautmann Wunsch. This chapter book tells the exciting story of the mission that placed Sojourner on the surface of Mars. (ages 9-12)

TAKE IT FURTHER

Payload Delivery

Can you design a delivery system that could drop food supplies into a cargo drop box on Mars? Let's find out! Your job today is to engineer a paper cup to carry a marble that will be released by a string in such a way that the marble hits a target on the floor. Does that sound difficult? It's really not!

You will need:
A paper or styrofoam cup, modified to deliver your payload
A travel line (can be string or fishing line)
A paper clip
A marble
String (to tip the cup)
Masking tape
A target affixed to the door

First, set up the travel line. It should be suspended in an incline above the floor. Make sure your cup will slide down it freely. Next, position a target on the floor underneath the travel line. Unfold the paper clip. Now, using tape, make a handle for the cup similar to that of a bucket handle. Attach the paper clip to the handle. Attach a release string to the bottom of the cup and the travel line's highest support. Your job is to modify the cup so that it can release the marble onto the target accurately and consistently. One way to do it is pictured below. Try it and then see if you can come up with a better design. Observe the results so that you can modify your delivery system to change the outcome. Have fun making your deliveries!

If your delivery system is unsuccessful, try these suggestions:
- mount a delivery platform on the outside of the cup
- cut a door in the cup
- add some rolled up tape in the bottom of the cup as runners to guide the marble out
- adjust the length of your release line if the marble does not hit the target
- change the angle of the cup if the marble does not stay on the platform or in the cup
- experiment with different targets: a sheet of paper, a box, a small box, a toy truck's cargo bay

Images and Experiment Courtesy of Cindy Frick

...the stars will fall from the sky, and the heavenly bodies will be shaken.' Mark 13:25

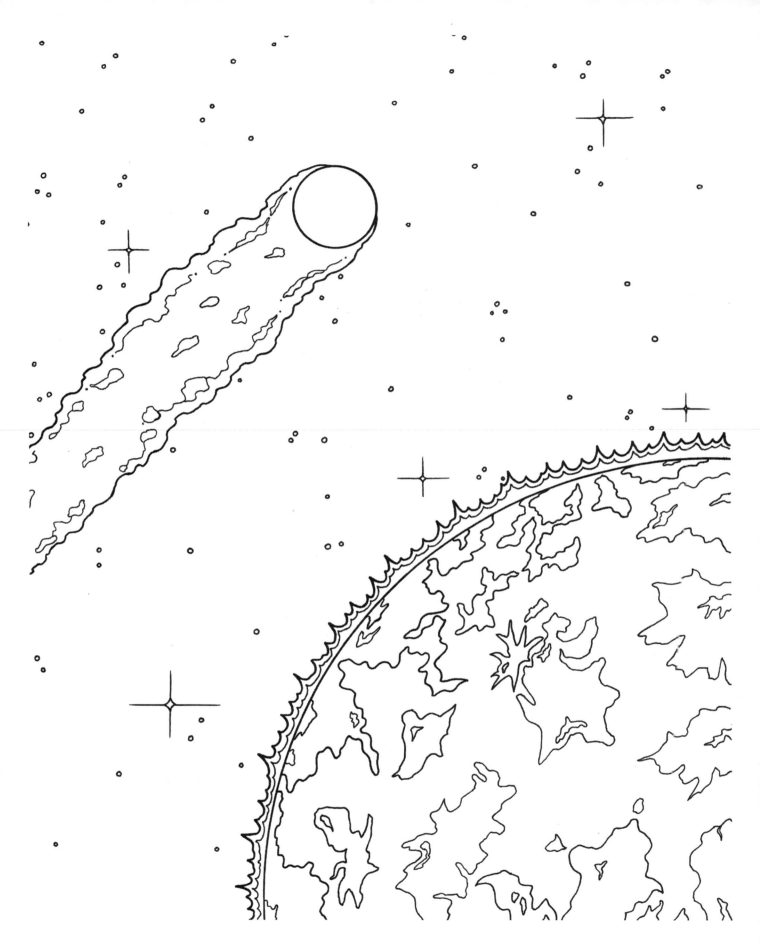

Great is our Lord and abundant in strength; His understanding is infinite.
Psalm 147:5

Fascinating Facts about Space Rocks
Lesson 8

Comets

Meteors

Asteroids

Meteor Showers

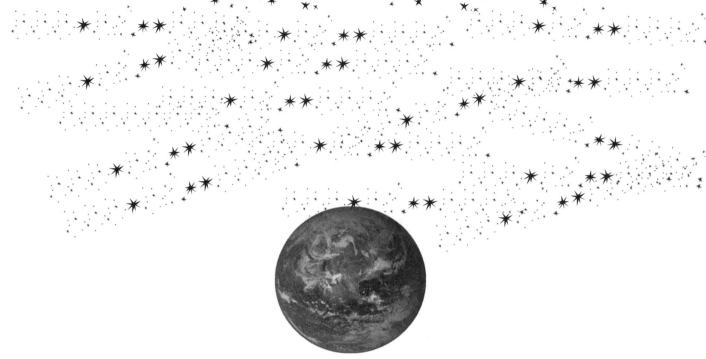

Shower Name	Approximate date to see	Name of comet	Number of shooting stars per hour
Quadrantids	January 3	Unknown	40
Pi Puppids	April 5	Unknown	40
Lyrids	April 22	Comet Thatcher	15
Eta Aquarids	May 5	Comet Halley	20
Delta Aquarids	July 30	Unknown	20
Perseids*	August 12	Comet Swift-Tuttle	50
Orionids	October 22	Comet Halley	25
Taurids	November 4	Comet Encke	15
Leonids	November 17	Comet Temple-Tuttle	15
Geminids*	December 14	Asteroid 3200 Phaethon	50
Ursids	December 23	Comet Tuttle	20

* Indicates large meteor showers

Trust in the LORD forever, for in GOD the LORD, we have an everlasting Rock.
Isaiah 26:4

Trust in the LORD forever, for in GOD the LORD, we have an everlasting Rock.
Isaiah 26:4

Space Rocks Vocabulary Story
Lesson 8

One day, I walked outside and right in my front yard was a giant hole that had not been there the day before. Inside the hole was a jagged rock. It was about 10 inches long and 5 inches wide. I realized that I discovered a _____ lying in my yard! I was so excited. Before this special rock was sitting in my yard, it had been in space. In space, it was called an _____ . But, somehow this special asteroid got in the way of the earth's orbit and was suddenly pulled into the earth's atmosphere. When it entered our atmosphere, it must have caught on fire. Many people would have thought it was a shooting star, but it was really a _____ . Sometimes a whole bunch of rocks and dust hit our atmosphere at once and create a lot of these shooting stars. We call this a _____ _____ .

I wonder if any more rocks will fall from the sky and hit my lawn. I know there are a lot of them up there circling the sun in the _____ _____ , which is right in between the _____ planets and the _____ planets. I wonder if there was once a planet there between Mars and Jupiter. Maybe that's why there are so many rocks right in that space between them.

I hope we never have a _____ hit our atmosphere. They are like huge, dirty snow balls. Some comets, like Halley's comet, come by the earth every 76 years. We call this a short period comet. If a comet takes longer than 200 years to orbit the sun, we call it a _____ _____ comet. Next time Halley's comet comes by the earth, it will be sometime after the year 2061. I will be _____ years-old! I hope I remember to watch for it.

Choose from the words below to fill in the blanks in the Venus Vocabulary Story.

Asteroid	Asteroid Belt	Meteorite
Meteor	Long Period	Comet
Inner	Meteor Shower	Outer

110

My Space Rocks Project
Lesson 8

What I did:

What I learned:

Space Rocks Minibook
Lesson 8

Paste your Space Rocks
Layered Book onto this
page.

TAKE IT FURTHER
LESSON 8

Solar System Model Options

Here are two optional activities to help you understand the distance between the planets. In both activities, you will need something to represent each planet and the sun. You can use the objects from the lesson on page 101, or something that is not to scale, such as marshmallows. In both activities, you will need a long hallway—such as you might find in a church.

Toilet Paper Model or Paper Clip Model

Using a roll of toilet paper or 392 paper clips, each sheet or clip represents 10,000,000 miles. Begin placing toilet paper sheets or paper clips from the sun to each planet in our solar system.

Planet	# of Sheets or Paper Clips from sun	# of Sheets or Paper Clips from previous planet
Mercury	4	4
Venus	7	3
Earth	10	3
Mars	15	5
Jupiter	52	37
Saturn	96	44
Uranus	192	96
Neptune	300	108

Dry Ice Comet Model

If you are feeling very adventurous, this activity is great fun for classes, giving everyone a real feel for comets. You can find dry ice in the yellow pages by searching under "ice." Make sure to bring an ice chest to transport the dry ice from the place of purchase to your freezer. *Remember that dry ice can be a dangerous chemical, so please use caution and only do this activity with adult supervision.*

You will need:
2 cups of water
2 cups of dry ice (frozen carbon dioxide)
2 spoonfuls of sand or dirt
A dash of ammonia
A large plastic mixing bowl
4 plastic garbage bags (put three of them inside of each other to make 1 three-ply bag)
Work gloves
A hammer
A large mixing spoon

Directions:
Line your mixing bowl with a garbage bag.
Pour in 2 cups of water.
Add sand or dirt and stir.
Add ammonia.
Add corn syrup and stir.
Put on your gloves.
Place the dry ice in the three-ply bag.
Pound and crush the dry ice using the hammer.
Add the dry ice to the mixing bowl, while stirring.
Stir until the mixture is frozen.
Use the bag in the bowl to shape the materials into a ball.
Unwrap the ball once it is frozen. This is your comet.

Model of Solar System with Asteroid Belt

On a sheet of construction paper, draw our solar system with colored markers, leaving a space for the asteroid belt. After drawing all the planets, use glue to trace the path of the asteroid belt. Sprinkle sand, salt, pepper, or glitter over the page so that it sticks to the glue, simulating the asteroid belt.

Cartoon of Exploded Planet

Create a cartoon of the events that led up to and culminated in the alleged explosion of the planet between Mars and Jupiter.

He set them in place for ever and ever; he gave a decree that will never pass away.　　　　　　　　　　　　　　　　Psalm 148:6

Thus the heavens and the earth were completed in all their vast array.
Genesis 2:1

Fascinating Facts
about
Jupiter
Lesson 9

For You light my lamp; the LORD my God illumines my darkness.

Psalm 18:28

COPYWORK

For You light my lamp; the LORD my God illumines my darkness.
                                        Psalm 18:28

Jupiter Vocabulary
Lesson 9

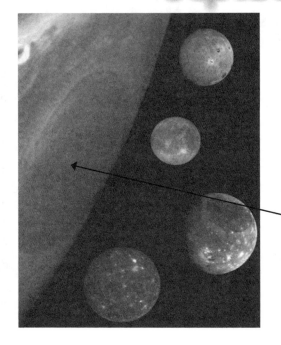

_____
- - - - - - - - - - -

_____
- - - - - - - - - - -

_____
- - - - - - - - - - -

_____
- - - - - - - - - - -

Choose from the words below to label the images of Jupiter.

Great Red Spot   **Galileo**
Jupiter   **Galilean Moons**

My Jupiter Project
Lesson 9

What I did:

What I learned:

JUPITER MINIBOOK
LESSON 9

Paste your Jupiter Shutter
Book onto this page.

TAKE IT FURTHER
LESSON 9

Planet Size Activity

It's been said that all the planets could fit inside Jupiter. You could do this experiment to see if that is true. You will need salt dough (or clay), a metric ruler, and the chart below to create a small scale model of each planet.

You can make your own salt dough using the recipe below.

Ingredients:
4 cups flour
1 cup salt
1-1/2 cups hot water
2 teaspoons vegetable oil

Mix the salt and flour together. While stirring, gradually add the water until the dough becomes elastic. Mix in the oil. If your dough is sticky, slowly add more flour. If it turns out too crumbly, add more water. Knead the dough until it's a nice consistency.

The numbers in the chart below compare the size of the planets to earth. Create the planets by forming balls. Measure each planet's diameter in centimeters by placing them against a ruler to make sure each one is the correct size. For example, Earth is 1.0 centimeters, and Mercury would be .38 centimeters or 3.8 millimeters.

Mercury	Venus	Earth	Mars	Jupiter	Saturn	Uranus	Neptune	Pluto
0.38	0.95	1.0	0.53	11.21	9.46	4.01	3.88	0.18

After you have created each one, take all the planets, except Jupiter, and roll them together to see how big of a planet they make. Are the combined planets bigger or smaller than Jupiter? Could they fit inside Jupiter?

Book Suggestions

Max Goes to Jupiter: a Science Adventure with Max the Dog by Jeffrey Bennet, Nick Schneider and Erica Ellingson. This picture book follows the adventures of Max, a friendly Rottweiler that travels through space with a small human crew of astronauts to visit the moons of Jupiter. (ages 4-8)
The Largest Planet: Jupiter by Nancy Loewen. The simple text and bold illustrations of this picture book introduce readers to Jupiter, the planet with the Great Red Spot. (ages 6-9)
Jupiter by Elaine Landau. Using lively narrative and photographs this book describes the characteristics of Jupiter and the history of space missions to the planet.
A Look at Jupiter by Ray Spangenburg and Kit Moser. This book's engaging text and photographs describe in depth man's earliest ideas about and observations of the planet Jupiter. (ages 10+)

The heavens declare the glory of God; the skies proclaim the work of his hands.

Psalm 19:1

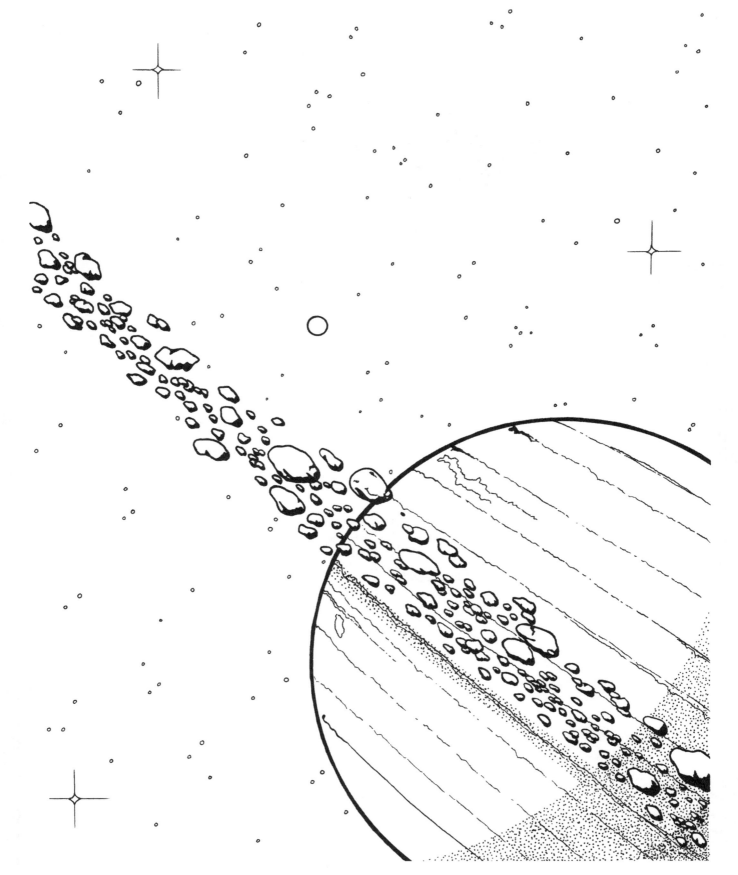

You alone are the LORD. You made the heavens, even the highest heavens, and all their starry host, the earth and all that is on it, the seas and all that is in them. You give life to everything, and the multitudes of heaven worship you.
Nehemiah 9:6

Fascinating Facts
About
Saturn
Lesson 10

Fascinating Facts about Saturn

Lesson 10

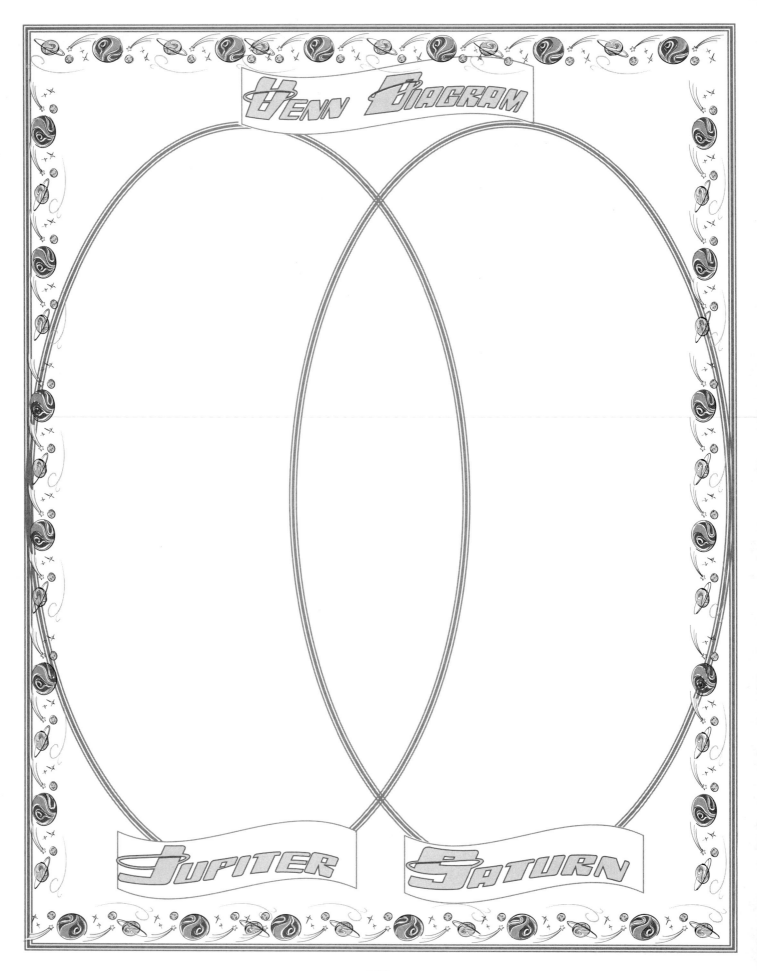

Vocabulary Crossword
Lesson 10

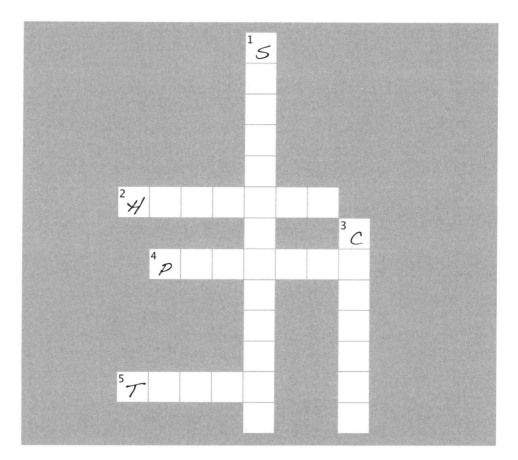

HUYGENS
SHEPHERD MOONS
TITAN

PANDORA
CASSINI

Across

2. The name of the man that discovered that Saturn's handles were actually rings around the planet. It's also the name of the probe that will collect and test samples of dirt from Titan, Saturn's moon.
4. One of Saturn's shepherd moons.
5. Saturn's largest moon. It is close to the size of Mercury.

Down

1. What we call Saturn's special moons that keep the rings on Saturn from spreading out too far.
3. The name of the mission and the spacecraft that traveled to Saturn.

He counts the number of the stars;
He gives names to all of them.
Psalm 147:4

He counts the number of the stars;
He gives names to all of them.
Psalm 147:4

My Saturn Project
Lesson 10

What I did:

What I learned:

SATURN MINIBOOK
LESSON 10

Paste your Saturn Pocket Book
onto this page.

TAKE IT FURTHER
LESSON 10

Make a Model of Saturn

You can make a beautiful model of Saturn that hangs from the ceiling.

You will need:
A yellow marker
A 2-3 inch Styrofoam ball
An unwanted CD
Glue
A paper clip
String
A tack
Glitter (gold and silver preferred)

Directions:
Begin by coloring the Styrofoam ball yellow with the marker. Cut the Styrofoam ball in half. Put glue on the flat insides of both halves. Place the CD inside these halves and put the ball back together with the CD in the center, resembling Saturn's rings. Now, glue the glitter on the ball in stripes to create the colors of Saturn. Last, insert the paper clip in the top of the ball. Attach string to the paper clip and hang the ball from the ceiling by tacking the other end of the string to the ceiling.

Book and DVD Suggestions

Saturn for My Birthday by John McGranaghan. In this delightful picture book, a young boy wants Saturn and its forty-seven moons for his birthday! (ages 4-8)

Ringed Giant: Saturn (Amazing Science) by Nancy Loewen. The simple but informative text and bold illustrations introduce young readers to the planet with nine rings. (ages 6-9)

Saturn (A True Book) by Elaine Landau. This book's lively narrative and liberal use of photographs describe the physical characteristics and history of space missions to Saturn. (ages 9-12)

A Look at Saturn (Out of This World) by Ray Spangenburg and Kit Moser. This book's photographs and engaging text describe in depth man's earliest ideas about and observations of the planet Saturn.

Bill Nye the Science Guy: Space Exploration distributed by Disney Educational Productions: 2004. Zaney Bill Nye meets NASA scientist Dr. Linda Horn while she's working on Cassini, the spacecraft destined for Saturn. (DVD for all ages)

"O LORD Almighty, God of Israel, enthroned between the cherubim, you alone are God over all the kingdoms of the earth. You have made heaven and earth." Isaiah 37:16

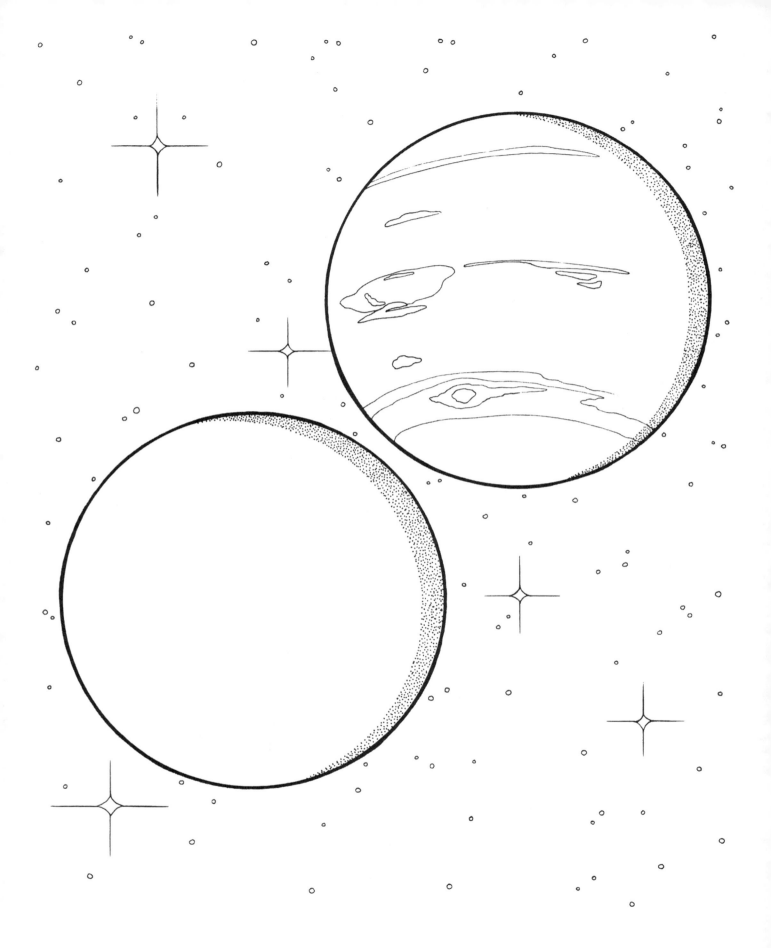

He wraps himself in light as with a garment; he stretches out the heavens like a tent.
Psalm 104:2

Fascinating Facts
About
Uranus and Neptune
Lesson 11

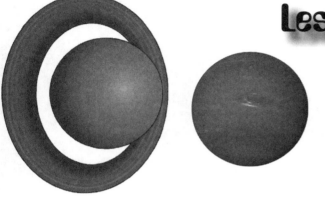

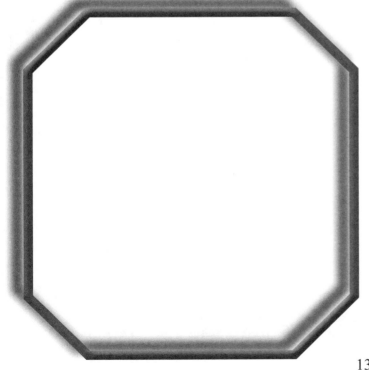

The Discovery of Uranus
Lesson 11

NEPTUNE & URANUS MATCH UP
LESSON 11

Cut out the vocabulary terms below and glue them inside the correct circles on the following pages.

350 DEGREES BELOW ZERO	TRITON	OPHELIA
RINGS	8TH PLANET	WILLIAM & CAROLINE HERSCHEL
17 HOUR DAY	16 HOUR DAY	164 YEAR ORBIT
7TH PLANET	DIAMOND RAIN	METHANE ICE
84 YEAR ORBIT AROUND THE SUN	WOBBLY WAGON WHEEL	BRIGHT BLUE COLOR
BLUE-GREEN COLOR	GEYSERS	GREAT DARK SPOT

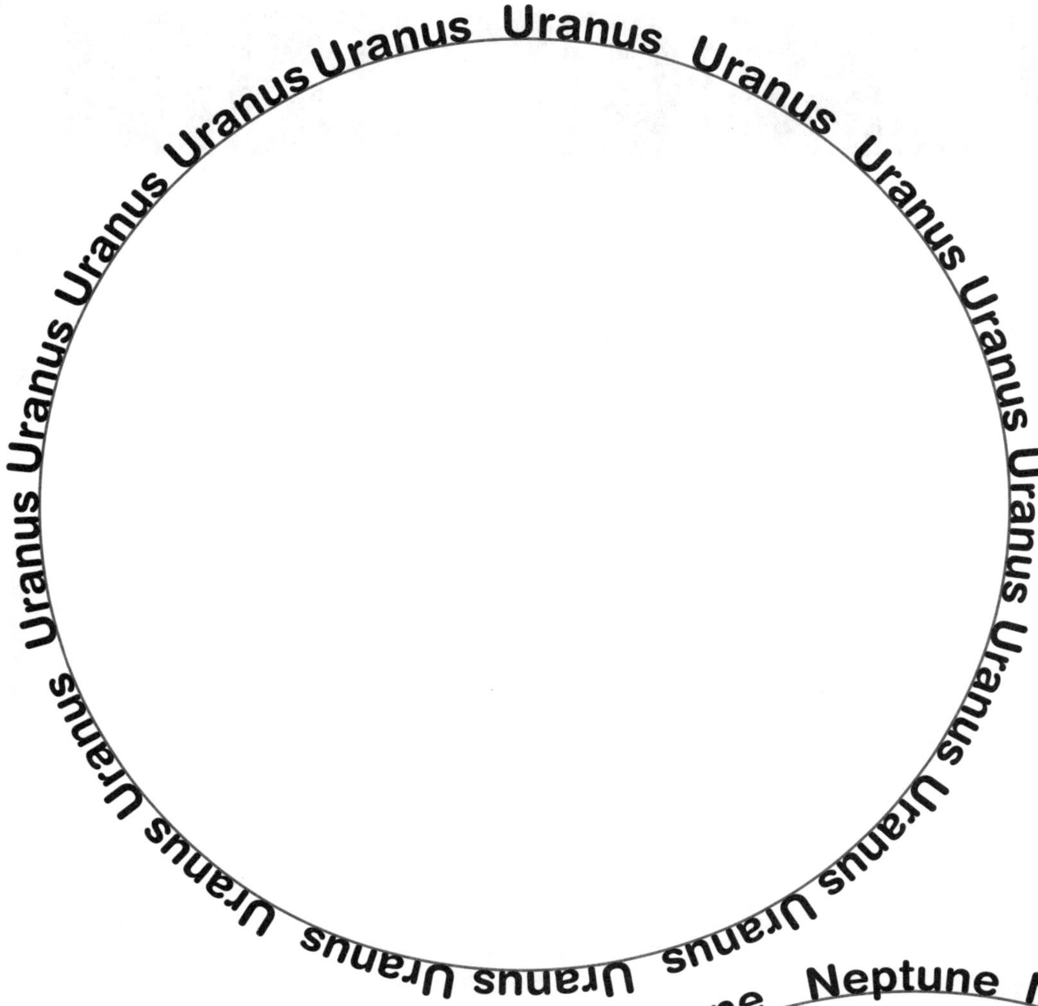
Uranus

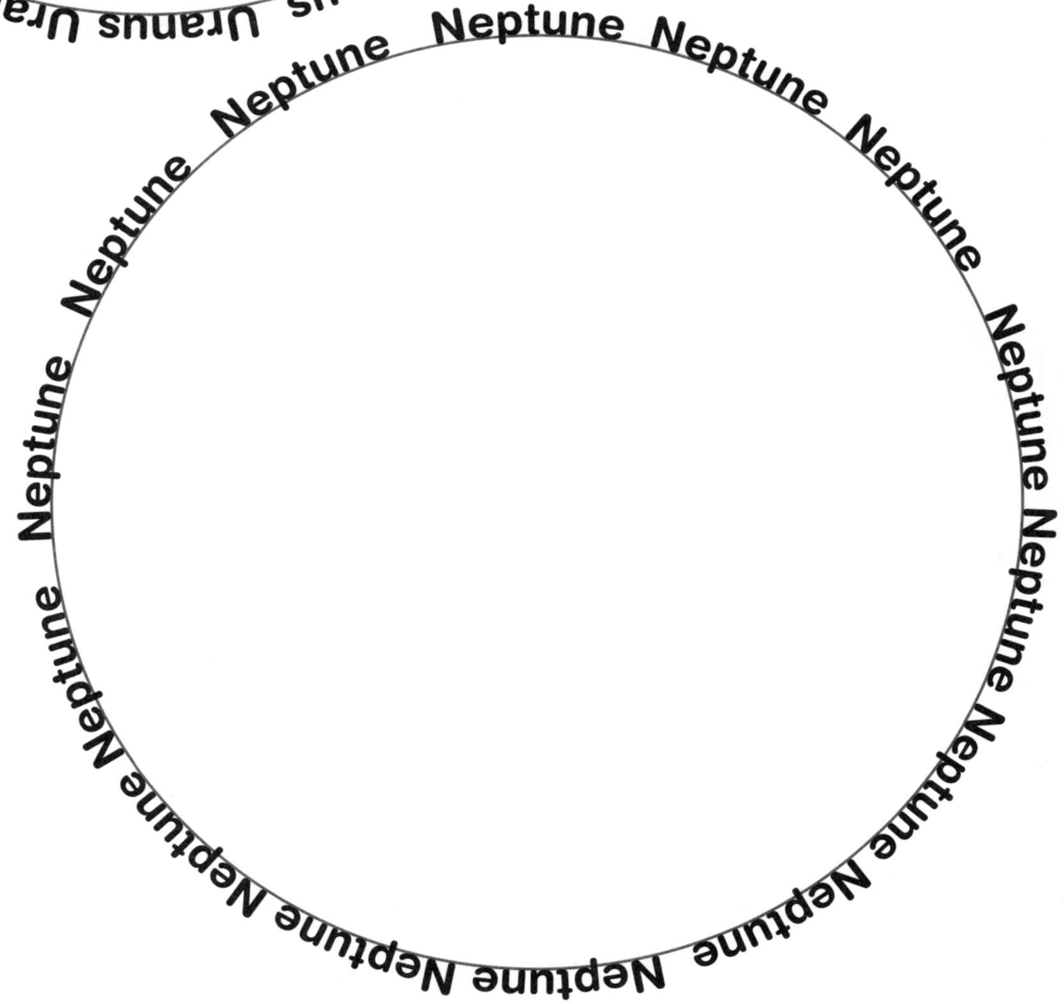
Neptune

For as the heavens are higher than the earth, so are My ways higher than your ways, and My thoughts than your thoughts.

Isaiah 55:9

For as the heavens are higher than the earth, so are My ways higher than your ways, and My thoughts than your thoughts.
Isaiah 55:9

My Uranus and Neptune Project
Lesson 11

What I did:

What I learned:

Uranus and Neptune Minibooks
Lesson 11

Paste your Uranus and
Neptune Pop-up Books onto
this page.

TAKE IT FURTHER
LESSON 11

Satellite Graph

You can create a Satellite Graph and record the number of satellites each planet has. Simply create a graph, then chart each planet and their satellites. This will help you to visually demonstrate the difference in the number of satellites, as well as reinforce what you have learned.

Planet Mystery Questions

Now that you have learned about all the planets, create a mystery question game to be played with your siblings or friends. Each one think up different mystery planet clues—for example, "I have rust," or "I'm made of methane and have rings." Put the questions on index cards and play a game to see how many each of you can get correct.

Book Suggestions

Farthest From the Sun: Neptune (Amazing Science) by Nancy Loewen. This picture book's informative text and bold illustrations introduce young readers to Neptune, the eighth planet from the sun. (ages 6-9)

The Sideways Planet: Uranus (Amazing Science) by Nancy Loewen. This picture book's informative text and bold illustrations introduce young readers to Uranus, the seventh planet from the sun. (ages 6-9)

Neptune (A True Book) by Elaine Landau. This book's lively text and photographs describe the physical characteristics of Neptune and the history of the space missions to the planet. (ages 9-12)

Uranus (A True Book) by Elaine Landau. This book's lively text and photographs describe the physical characteristics of Uranus and the history of the space missions to the planet. (ages 9-12)

A Look at Neptune (Out of This World) by John Tabak. With engaging text and photographs, this chapter book describes in depth man's earliest ideas about and observations of Neptune. (ages 10+)

A Look at Uranus (Out of This World) by Salvatori Tocci. With engaging text and photographs, this chapter book describes in depth man's earliest ideas about and observations of Uranus. (ages 10+)

This is what God the LORD says--he who created the heavens and stretched them out, who spread out the earth and all that comes out of it, who gives breath to its people, and life to those who walk on it:

Isaiah 42:5

Fix these words of mine in your hearts and minds; tie them as symbols on your hands and bind them on your foreheads. Teach them to your children, talking about them when you sit at home and when you walk along the road, when you lie down and when you get up. Write them on the door frames of your houses and on your gates, so that your days and the days of your children may be many in the land that the Lord swore to give your forefathers, as many as the days that the heavens are above the earth.
Deuteronomy 11:18-21

Fascinating Facts
About
Pluto and the Kuiper Belt
Lesson 12

Pluto Debate
Lesson 12

1st Point

2nd Point

My Hypothesis

3rd Point

4th Point

5th Point

Vocabulary Puzzle Game

NEW HORIZONS

HYPOTHESIS

KUIPER BELT

The name of Pluto's moon.

An object in the Kuiper belt that is about half the size of Pluto. It was discovered in 2002.

Objects in the Kuiper belt that look like Pluto.

Cut out each puzzle piece on this page and the next. Match each vocabulary word puzzle piece to the correct definition puzzle piece. Cut out this rectangle and glue it to your Puzzle Page along the bottom and side edges to create a pocket. After you've played your Vocabulary Puzzle Game a few times, place all your puzzle pieces in the pocket for safe keeping.

-Glue along this edge-

-Glue along this edge-

-Glue along this edge-

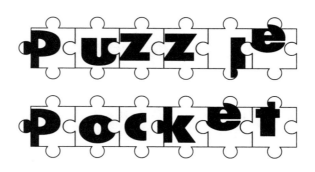

CHARON

PLUTINOS

A ring of objects that orbit the sun outside our Solar System.

A good guess based on the facts.

The unmanned space craft that will reach Pluto in the year 2015.

QUAOAR

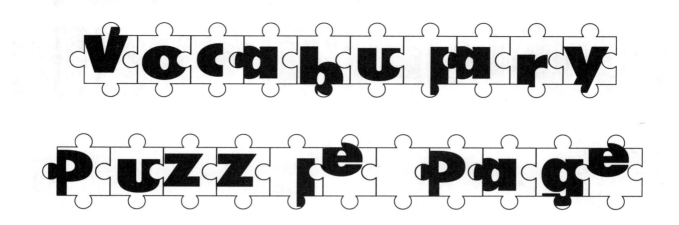

Paste your puzzle pocket here.

"Those who have insight will shine brightly like the brightness of the expanse of heaven, and those who lead the many to righteousness, like the stars forever and ever."

Daniel 12:3

"Those who have insight will shine brightly like the brightness of the expanse of heaven, and those who lead the many to righteousness, like the stars forever and ever."

Daniel 12:3

My Pluto & Kuiper Belt Project
Lesson 12

What I did:

What I learned:

Pluto & the Kuiper Belt Minibooks
Lesson 12

Paste your Pluto Debate and
Kuiper Belt Minibooks onto
this page.

TAKE IT FURTHER
Lesson 12

Pluto in History

It takes Pluto 248 years to orbit the earth. The last time Pluto was in the position it is in today, what was going on with the earth? Research what was happening on the earth 248 years ago. You can create a paper or poster of all the things that were going on. Include what people wore, where they lived, and how they ate. If you have any information about your personal ancestors, this might be a great time to discover what their lives were like 248 years ago.

Make a Construction Paper Model of the Solar System

Using the image found on page 132 of *Exploring Creation with Astronomy* as a guide, make a model of the solar system with Pluto and the Kuiper belt represented.

Create your model on construction paper, using yarn or string to represent the orbital path of the planets, Pluto, and the Kuiper belt. You might use beans or beads to represent each planet. Salt rocks or glitter might be a good thing to represent the Kuiper belt. Don't forget the asteroid belt in your model.

Book Suggestions

Why Isn't Pluto a Planet? A Book About Planets by Steve Kortenkamp. Using simple text and bold illustrations, this picture book answers some of the questions young children have about our solar system. (ages 4-8)

The Planet Hunter: The Story of What Happened to Pluto by Elizabeth Rusch. This picture book tells the story of astronomer Mike Brown, detailing his childhood fascination with our solar system and his amazing research and discoveries that changed the status of Pluto. (ages 4-8)

Beyond Pluto: The Final Frontier in Space by Elaine Landau. Using lively text and beautiful photographs, this book describes the history of the observation and discovery of the Kuiper belt. (ages 9-12)

Pluto: From Planet to Dwarf by Elaine Landau. This book describes the history and discovery of Pluto and how astronomers used modern technology to determine it is a dwarf planet. (ages 9-12)

When is a Planet Not a Planet? by Elaine Scott. This delightful book shares the history of astronomical discovery and debate and examines the technology used to demote Pluto from the status of planet to dwarf planet.

Those who are wise will shine like the brightness of the heavens, and those who lead many to righteousness, like the stars for ever and ever.   Daniel 12:3

...so that you may become blameless and pure, children of God without fault in a crooked and depraved generation, in which you shine like stars in the universe.   Philippians 2:15

Fascinating Facts
About
Stars & Galaxies
Lesson 13

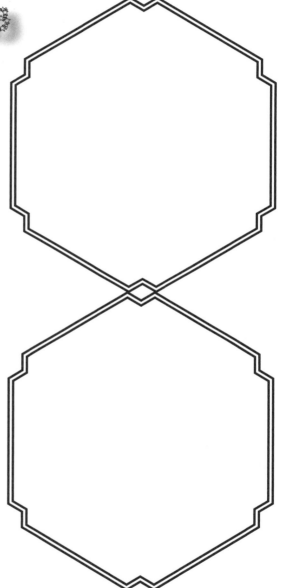

Star Temperature Mnemonic

O	B	A	F	G	K	M

Fascinating Facts
about
Stars & Galaxies
Lesson 13

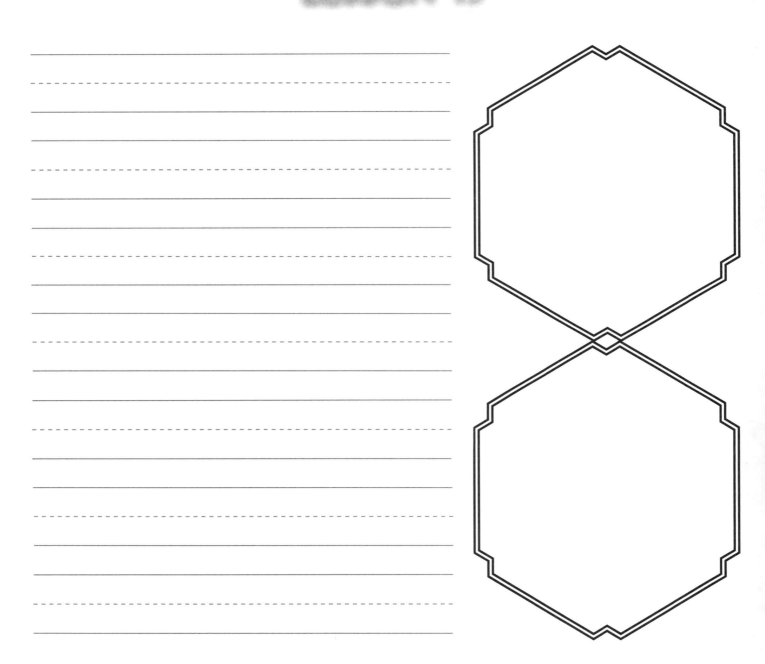

CONSTELLATIONS IN MY AREA
LESSON 13

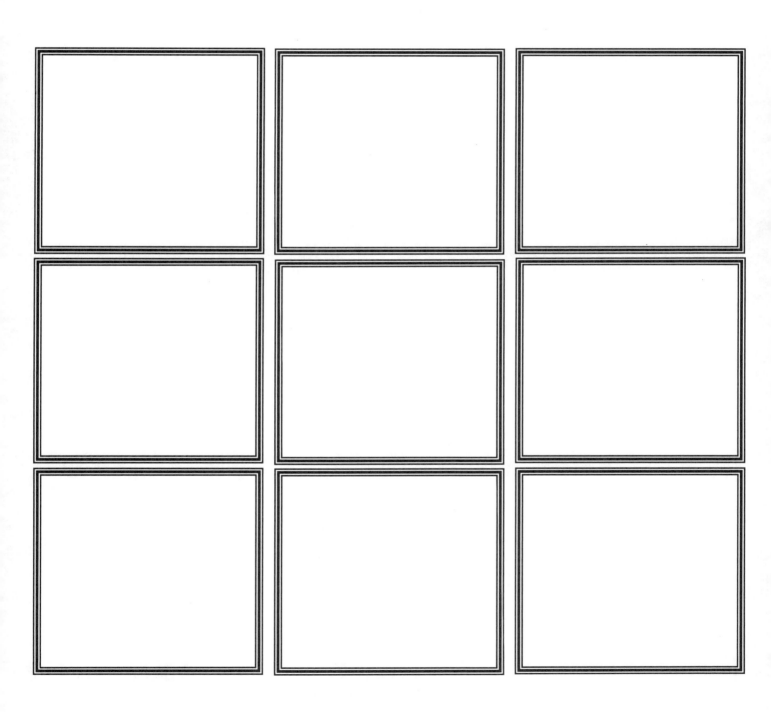

165

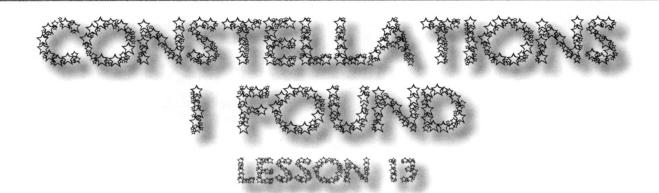

Stars & Galaxies Identification
Lesson 13

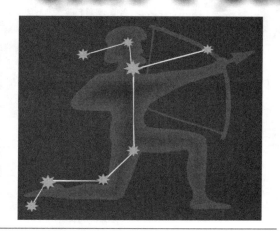

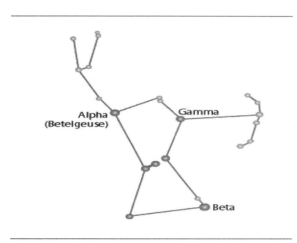

Choose from the words below to label the images.

Big Dipper **Galaxy** **Constellation**
Orion **Little Dipper** **North Star**

The heavens are telling of the glory of God; and their expanse is declaring the work of His hands. Day to day pours forth speech, and night to night reveals knowledge.

Psalm 19:1-2

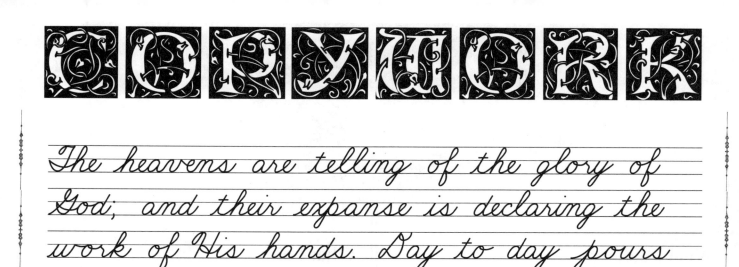

The heavens are telling of the glory of God, and their expanse is declaring the work of His hands. Day to day pours forth speech, and night to night reveals knowledge.

Psalm 19:1-2

MY STARS & GALAXIES PROJECTS
LESSON 13

What I did:

What I did:

What I learned:

What I learned:

STARS & GALAXIES MINIBOOK
LESSON 13

Paste your Stars & Galaxies
Fan onto this page.

TAKE IT FURTHER
LESSON 13

Ancient Eye Test

The middle star in the Big Dipper's handle is a star called Mizar. There is a star lined up with Mizar named Alcor. They appear so close together that, to some, they can look like one star. In ancient days, these stars were sometimes used as an eye test. If you could see two separate stars, your eyesight was good; if not, you had poor eyesight.

On a clear night, go outside and look at the middle star in the handle of the Big Dipper. Is it one star, or can you see two? Look through some binoculars to get your answer.

Create a Constellation Myth

All of the constellations come with stories. The Ancient Greeks had stories for the constellations, as did the Ancient Indians, Native Americans and Chinese. Often these stories were quite similar to one another, and the constellations represented the same object or person, even though they don't look like the person or object they supposedly represent. These stories surrounded ancient mythical beings, people, animals and situations. For example, the story of Ursa Major and Ursa Minor revolves around the mythical heroes Jupiter and his wife, who turn a woman and her son into bears. If you feel up to it, research some of the constellation stories before doing this activity.

Choose a group of stars in the sky to create your own constellation. You can give it a name and make up a story that tells why the constellation is there in the sky and what it is doing.

Make a Star Wheel

A star wheel, or planisphere, is a device that helps you locate the stars in the sky during the different seasons. On the Internet, there are many sites that have printable star wheels you can make. Do a search for: "Make a Star Wheel" and you should find several to print and put together.

Books Suggestions

The Stars: A New Way to See Them by H.A. Rey. This classic picture book provides a bulk of information on stars, planets and constellations. (ages 9-12)
Find the Constellations by H.A. Rey. This classic book helps readers identify the constellations of the night sky. Many delightful yet clear illustrations are included. (ages 9-12)
Constellations by F.S. Kim. Using lively narrative and photographs, this book describes the history of the myths of constellations and helps readers identify them in the night sky. (ages 9-12)
Super Stars: The Biggest, Brightest and Most Explosive Stars in the Milky Way by David Aguilar. This book uses lively text and brilliant photographs to explore fifteen of the best known stars of our solar system, and also other interesting stellar objects. (ages 9-12)
Black Holes by Ker Than. This book's lively narrative and use of photographs explain how stars become black holes, what is inside them, and how scientists study them. (ages 9-12)
The Mysterious Universe: Supernovae, Dark Energy, and Black Holes by Ellen B. Jackson. This book follows astronomers as they hunt for supernovae and large-scale astronomical phenomena. (ages 9-12)
The Life and Death of Stars by Ray Spangenburg and Kit Moser. This chapter book describes man's earliest observations and discoveries of the stars, with an emphasis on space missions taken to gather more information about them. (ages 10+)

For we are God's workmanship, created in Christ Jesus to do good works, which God prepared in advance for us to do.
Philippians 2:10

And we pray this in order that you may live a life worthy of the Lord and may please him in every way: bearing fruit in every good work, growing in the knowledge of God.   Colossians 1:10

Fascinating Facts
about
Space Travel
Lesson 14

Let's Visit the Planets!
Lesson 14

The day you leave the earth is your next birthday.	How old are you the day you leave?
You arrive on Mercury in 8 months.	How old are you when you get to Mercury? (your age + 8 months)
You arrive on Venus 4 months later.	How old are you when you reach Venus? (your last age + 4 months)
You arrive back on earth 4 months later and say hello to your family before you immediately set off for Mars.	How old are you when you reach earth? (your age + 4 months)
You arrive on Mars in 7 months.	How old are you when you reach Mars?
You arrive on Jupiter in 3 years and 11 months.	How old are you when you reach Jupiter?
You arrive on Saturn 4 years and 7 months later.	How old are you when you reach Saturn?
You arrive on Uranus 10 years and 2 months later.	How old are you when you reach Uranus?
You arrive on Neptune 11 years and 7 months later.	How old are you when you reach Neptune?
You arrive on Pluto 10 years later.	How old are you when you reach Pluto?
It will take 40 years and 10 months to make it back home.	How old will you be when you get back?

Vocabulary Puzzle Game

ARMSTRONG

COSMONAUT

SPUTNIK

The name of the space competition between the United States and Russia.

The man who studied how a person might get to the moon. Some called him "Moon Man."

Extravehicular activity, or a space walk.

Cut out each puzzle piece on this page and the next. Match each vocabulary word puzzle piece to the correct definition puzzle piece. Cut out this rectangle and glue it to your Puzzle Page along the bottom and side edges to create a pocket. After you've played your Vocabulary Puzzle Game a few times, place all your puzzle pieces in the pocket for safe keeping.

—Glue along this edge—
—Glue along this edge—
—Glue along this edge—

SPACE RACE

The first rocket that was launched into space. It was built by the Russians.

The last name of the first man to step on the moon.

EVA

What Russians call human scientists that travel to space.

GODDARD

179

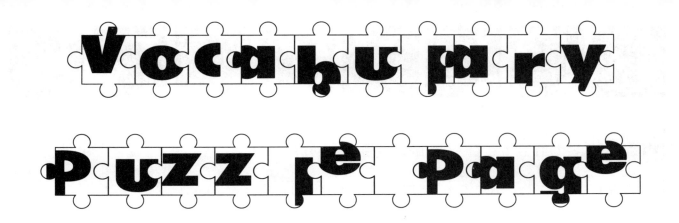

Paste your puzzle pocket here.

When I consider Your heavens, the work of Your fingers, the moon and the stars, which You have ordained; what is man that You take thought of him, and the son of man that You care for him?

Psalm 8:3-4

When I consider Your heavens, the work of Your fingers, the moon and the stars, which You have ordained; what is man that You take thought of him, and the son of man that You care for him?

Psalm 8:3-4

My Space Travel Project
Lesson 14

What I did:

What I learned:

Space Travel Minibook
Lesson 14

Paste your Space Travel Lift
and Learn onto this page.

Planets Review Minibook
Lesson 14

Paste your Solar System
Review Wheel onto this page.

TAKE IT FURTHER
LESSON 14

Visit a Planetarium

This might be a nice time to schedule a trip to a planetarium. A planetarium is a large, usually dome-shaped theater built to educate the public on the night's sky and astronomy. These are exciting and interesting theaters and you will enjoy the show; however, be forewarned that evolutionary content may be presented.

Many universities have a planetarium, as do most large cities. Look up the word "planetarium" on the Internet and find one that is near to where you live. You may want to call to schedule your visit.

Enter a Science Fair

If you have never participated in a science fair, this is a worthwhile activity. There are numerous resources to help you choose a science fair project. If you wish to do one based on astronomy, you can use one of the following books:

Janice Van Cleave's A+ Projects in Astronomy by Janice Van Cleave
Science Fair Projects: Flight, Space & Astronomy by Robert L. Bonnet
Astronomy and Space by Kelly Milner Halls

Book Suggestions

Can You Hear a Shout in Space? Questions and Answers about Space Exploration by Melvin and Gilda Berger. This richly illustrated book attempts to answer many of the questions children have about space and its exploration. (ages 6-9)
Look to the Stars by Buzz Aldrin. Written by a former prominent NASA astronaut, the vibrant illustrations and engaging text of this picture book take the reader on a journey through the history of air and space travel. (ages 6-9)
The Hubble Space Telescope by Diane and Paul Sipiera. This book describes how the Hubble Space Telescope was placed into orbit to provide us with information about outer space. (ages 9-12)
Rocket Man: The Mercury Adventure of John Glenn by Ruth Ashby. This dramatic story tells of John Glenn's near-disasterous mission in Friendship 7. (ages 9-12)
Space Exploration (DK Eyewitness) by Carol Stott. This comprehensive resource covers in depth all aspects of space exploration, including its rich history and exciting future. (ages 9-12)
Exploring the Solar System: a History with 22 activities by Mary Kay Carson. This comprehensive book describes the rich history of space exploration. (ages 10+)
The Hubble Space Telescope by Ray Spangenburg and Kit Moser. This chapter book uses engaging text and photographs to detail the history of telescopes and the vision and journey of the Hubble's successful launch. (ages 10+)
Mission Control, This is Apollo: the Story of the First Voyages to the Moon by Andrew Chaikin. This book recounts the complete history of man's pursuit of moon exploration, from Mercury to Apollo 17 and beyond.
NASA by Salvatore Tocci. The history of NASA and its missions into outer space are described in this chapter book. (ages10+)

Answers to the Vocabulary Activities

Vocabulary Crossword
Lesson 1

	1	2	3	4	5	6	7	8
1	C	O	P	E	R	N	I	C	U	S

Across:
1. COPERNICUS
3. GALILEO
5. INSTINCT
7. UNIVERSE

Down:
2. STONEHENGE
4. GRAVITY
6. NCASA

189

SUN VOCABULARY
LESSON 2

ECLIPSE
When the moon is between the earth and the sun, completely covering the sun, causing the moon to cast a shadow upon the earth.

ORBIT
The path an object or planet follows as it moves around the sun.

ATMOSPHERE
The layers of gases and mist that cover a planet, such as the earth.

THERMONUCLEAR FUSION
Nuclear reactions (controlled explosions) that happen on the sun that give it energy and power.

SOLAR FLARES
Large fires that burst out millions of miles from the sun, releasing energy into the solar system.

SUNSPOTS
Dark patches on the sun which are cooler than the rest of the sun.

AURORAS
Natural, colorful light displays in the sky that can be caused by solar flares.

ROTATE
To turn around a fixed point, the way a top spins in place or a planet spins on its axis.

Vocabulary Puzzle Game Answers Lesson 3

TRANSIT
This word means to pass over. We use this word to describe when Mercury passes between the earh and the sun.

GASEOUS
Planets that are not solid, but instead are made of gas.

CRATER
A dent on the surface of a planet or moon caused by the impact of an asteroid.

ASTEROIDS
Rocks that orbit in space, sometimes crashing into planets and other satellites.

TERRESTRIAL
Planets that are "earth-like," having a solid surface upon which you can stand.

UNMANNED
A spacecraft that travels without a person inside.

Venus Vocabulary Story
Lesson 4

I looked up in the sky right before the sun went down and saw a little star brightly shining in the sky near the horizon. I pointed to it and said, "Look! The first star of the evening." My mom said, "That's not a star. That's the planet __Venus__. But many people call it the __Evening Star__ when they see it in the evening. It is very close to the sun, so we see it when the sun is coming up or going down.

My mom explained that Venus goes through phases, just like the moon. So, when we look at Venus through a telescope, it might be perfectly round, like a full moon. Or it might be in the shape of a __Crescent__, like when we see only a sliver of the moon. The planet doesn't actually change shapes, it's just the way the sun's light hits it.

A few weeks later, I woke up at the crack of dawn and looked out my window. Guess what I saw! I saw Venus again! It was shining so early in the morning. I remembered my mom told me that when Venus is seen early in the morning, we call it the __Morning Star__.

I wish I could travel in a spaceship to Venus and see what it looks like up close. My mom says that would be very dangerous because the surface of Venus has many volcanoes on it. The volcanoes spill out __Lava__, which is hot, liquid rock. When the liquid dries, it leaves really jagged rocks all over the place. So, Venus isn't as pretty as the earth, but I still think it would be exciting to see it in real life!

Vocabulary Lift the Flap
Lesson 5

The imaginary line or circle around the center of the earth.

A large magnetic field that surrounds the earth.

The layer of the earth just below the earth's crust.

The solid outermost layer of the earth.

The solid center of the earth.

The only planet in our solar system that can support life.

Moon Phases
Lesson 6

Full

Gibbous

Quarter

Crescent

Vocabulary Crossword
Lesson 7

Across:
4. DEIMOS
5. POLAR ICE CAPS
6. DRY ICE
7. ECOSYSTEM

Down:
1. PHOBOS
2. OLYMPUS MONS
3. PERMAFROST

Space Rocks Vocabulary Story
Lesson 8

One day, I walked outside and right in my front yard was a giant hole that had not been there the day before. Inside the hole was a jagged rock. It was about 10 inches long and 5 inches wide. I realized that I discovered a __Meteorite__ lying in my yard! I was so excited. Before this special rock was sitting in my yard, it had been in space. In space, it was called an __Asteroid__. But, somehow this special asteroid got in the way of the earth's orbit and was suddenly pulled into the earth's atmosphere. When it entered our atmosphere, it must have caught on fire. Many people would have thought it was a shooting star, but it was really a __Meteor__. Sometimes a whole bunch of rocks and dust hit our atmosphere at once and create a lot of these shooting stars. We call this a __Meteor Shower__.

I wonder if any more rocks will fall from the sky and hit my lawn. I know there are a lot of them up there circling the sun in the __Asteroid__ __Belt__, which is right in between the __Outer__ planets and the __Inner__ planets. I wonder if there was once a planet there between Mars and Jupiter. Maybe that's why there are so many rocks right in that space between them.

I hope we never have a __Comet__ hit our atmosphere. They are like huge dirty snow balls. Some comets, like Halley's comet, comes by the earth every 76 years. We call this a short period comet. If a comet takes longer than 200 years to orbit the sun, we call it a __Long__ __Period__ comet. Next time Halley's comet comes by the earth, it will be sometime after the year 2061. I will be _____ years-old! I hope I remember to watch for it.

Jupiter Vocabulary
Lesson 9

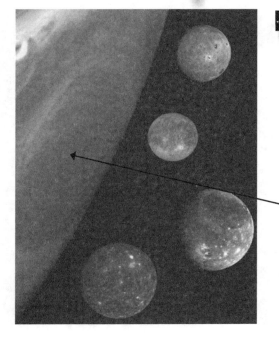

Great Red Spot

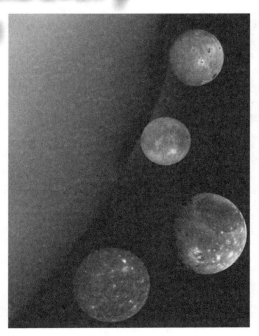

Galilean Moons

Galileo

Jupiter

Vocabulary Crossword
Lesson 10

Across:
2. HUYGENS
4. PANDORA
5. TITANS

Down:
1. SHEPHERDMOONS
3. CASSINI

198

NEPTUNE & URANUS MATCH UP
LESSON 11

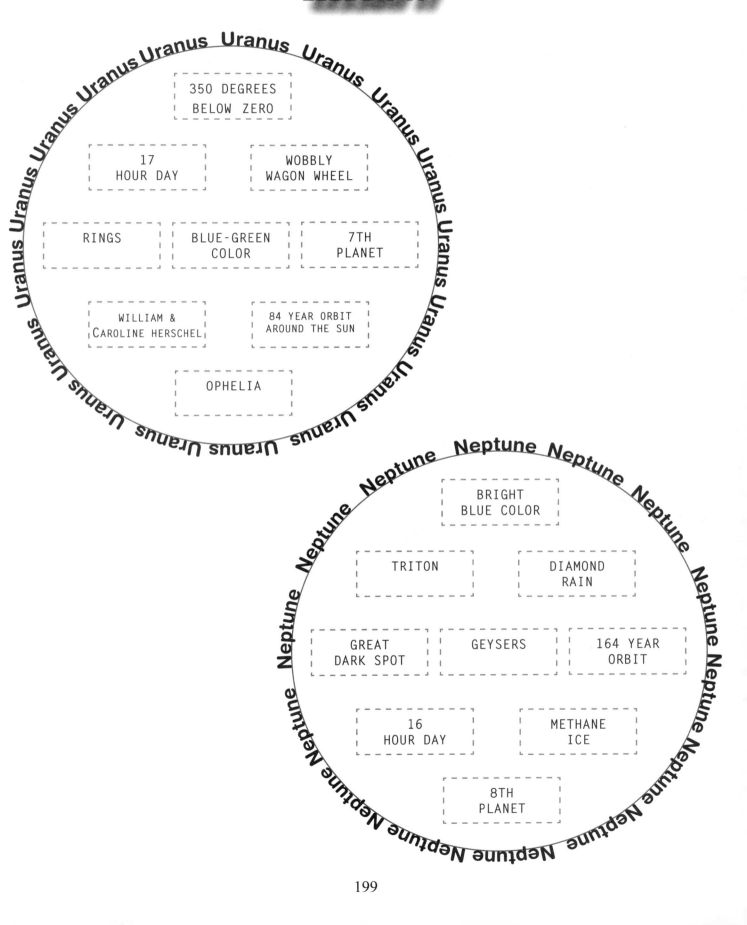

NEW HORIZONS

The unmanned space craft that will reach Pluto in the year 2015.

HYPOTHESIS

A good guess based on the facts.

PLUTINOS

Objects in the Kuiper belt that look like Pluto.

KUIPER BELT

A ring of comet-like objects that orbit the sun outside our solar system.

CHARON

The name of Pluto's moon.

QUAOAR

An object in the Kuiper belt that is about half the size of Pluto. It was discovered in 2002.

Stars & Galaxies Identification
Lesson 13

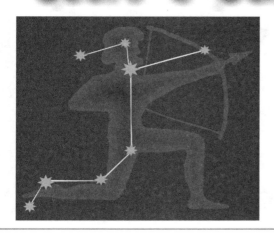
Orion

Little Dipper

Galaxy

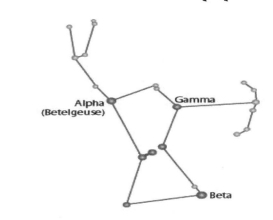
Constellation

North Star

Big Dipper

Vocabulary Puzzle Game Answers Lesson 14

ARMSTRONG

The last name of the first man to step on the moon.

COSMONAUT

What Russians call human scientists that travel to space.

EVA

Extravehicular activity, or a space walk.

SPUTNIK

The first rocket that was launched into space. It was built by the Russians.

SPACE RACE

The name of the space competition between the United States and Russia.

GODDARD

The man who studied how a person might get to the moon. Some called him "Moon Man."

My Astronomy Field Trip

Place: **Date:**

The purpose of this field trip:

What I saw/ did on this trip:

What I learned:

My favorite part:

My Astronomy Field Trip

Place: **Date:**

The purpose of this field trip:

What I saw/ did on this trip:

What I learned:

My favorite part:

My Astronomy Field Trip

Place: **Date:**

The purpose of this field trip:

What I saw/ did on this trip:

What I learned:

My favorite part:

My Astronomy Field Trip

Place: **Date:**

The purpose of this field trip:

What I saw/ did on this trip:

What I learned:

My favorite part:

My Astronomy Field Trip

Place: **Date:**

The purpose of this field trip:

What I saw/ did on this trip:

What I learned:

My favorite part:

My Astronomy Field Trip

Place: **Date:**

The purpose of this field trip:

What I saw/ did on this trip:

What I learned:

My favorite part:

CREATION CONFIRMATION MINIATURE BOOK

(Instructions on back)

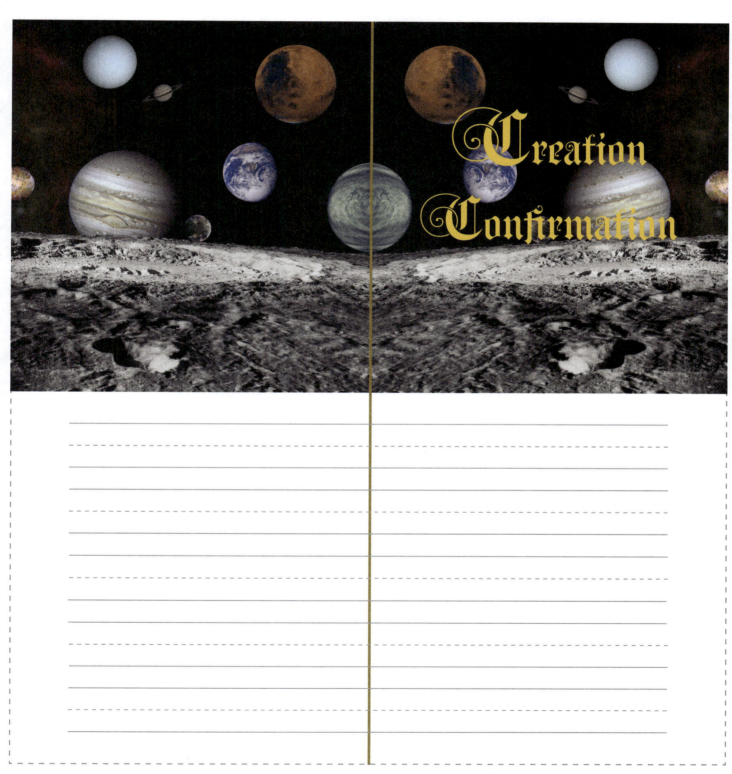

Creation Confirmation Miniature Book   A 1

It's important to remember all you've learned about God and Creation in this course. This Creation Confirmation Book will enable you to record and recall your learning.

Instructions:
1. Cut out the Creation Confirmation Book rectangles on pages A1 and A3 along the dotted lines. **Do not cut the gold fold lines!**
2. Fold the pages along the gold lines.
3. Place the pages inside the cover of the book.
4. Open the book to the middle and staple it along the center.
5. As you work through each lesson of the astronomy course, write down what you learn about God, the Bible and Creation.
6. Keep your Creation Confirmation Book inside your astronomy book as a book mark and a reminder to write down the things you learn about God.

Creation Confirmation Miniature Book   A 3

EXTRA MINIATURE BOOKS

Glue this side to your paste page.

If you would like to record any additional information not included in the other miniature books, here are a few extra miniature books for you to use. Paste them anywhere in your notebooking journal.

Instructions:

1. Cut out the three minibook covers along the dotted lines. **Do not cut the gold fold lines!**
2. Fold the covers in half along the gold fold lines.
3. Have fun recording all the interesting facts you learn about astronomy!

Glue this side to your paste page.

Glue this side to your paste page.

Extra Miniature Books   A 5

Extra Miniature Books

WHAT IS ASTRONOMY? MATCHBOOK

This is the matchbook cover that will hold all your square pages.

Instructions:

1. Cut out the matchbook cover along the dotted lines. **Do not cut the blue fold lines!**
2. Fold along the blue lines so that the large Solar System flap and the small flap face outward in the same direction.
3. Cut out all eight squares on this page and the next and fill in the information you learned about astronomy.
4. Lift the large flap and place all the pages you created under the small flap.
5. With the large cover flap open and your eight pages under the small flap, staple your matchbook on the white line that crosses the center of the small flap. This will hold all your pages inside. **Do not staple the cover closed!**
6. Fold the large flap down and tuck it into the small flap, like a matchbook.
7. Glue this side (with these words) onto your "Astronomy Minibooks" paste page *(NJ p. 18)*.

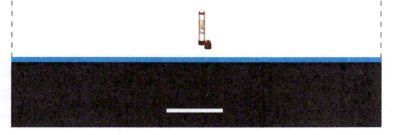

What is Astronomy? Matchbook: Lesson 1   A 7

Galileo

Nicolas Copernicus

NASA

Stonehenge

Constellations

Natural Satellite

What is Astronomy? Matchbook: Lesson 1

Yellow Circle

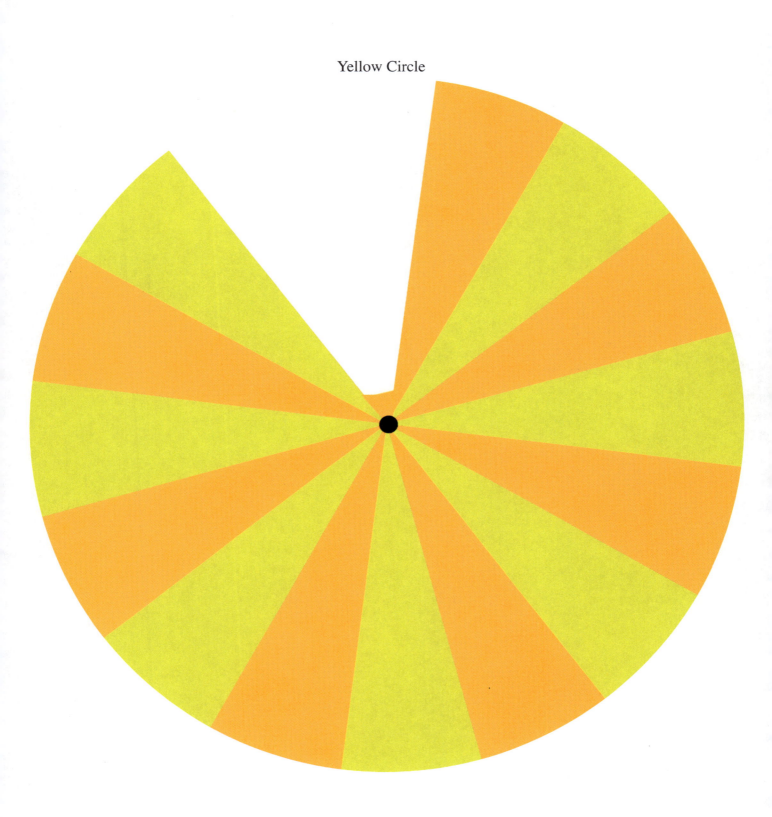

Sun Wheel: Lesson 2   A 11

Instructions:

1. Cut out the Fact Circle and the Yellow Circle. Be certain to cut out the white triangle in the Yellow Circle.
2. Place the Yellow Circle on top of the Fact Circle and insert a brass fastener in the center to secure the two circles.
3. Dab glue on the bottom of the Fact Circle and glue your Sun Wheel onto your "Sun Minibooks" paste page *(NJ p. 33)*.
4. In the empty triangle below each topic, write a fact you learned about that topic listed in the Fact Circle.
5. Turn the Yellow Circle around to reveal the different facts about each topic.

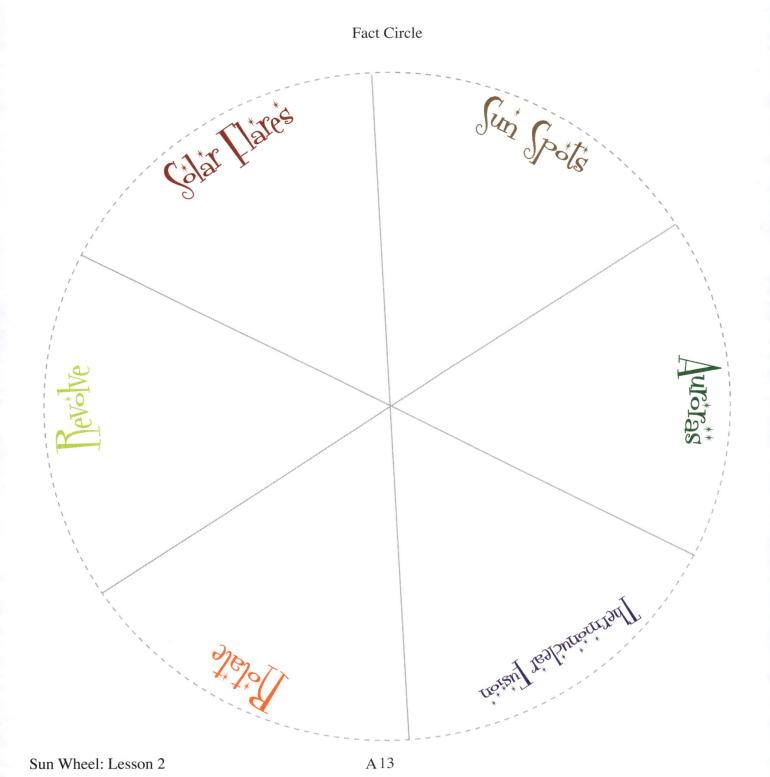

Fact Circle

Sun Wheel: Lesson 2   A 13

SUN MINIBOOKS

Rainbow Book Instructions:

1. Cut out the rainbow along the outside dotted lines and fold it in half so that the words appear on the outside.
2. Cut out the small circle in the center while it is folded in half (you will cut a semicircle, which will reveal a full circle when the rainbow opens).
3. Open the rainbow book and write what you learned about color on the inside.
4. Affix your Rainbow Book onto your "Sun Minibooks" paste page *(NJ p. 33)*.

Solar Eclipse Book Instructions:

1. Cut out the Solar Eclipse Book around the half-circles and along the dotted lines. **Do not cut the gold fold lines!**
2. Fold the yellow and black semi-circles inward along the gold lines.
3. Open the flaps and draw an illustration of a solar eclipse with the earth, moon and sun on the inside.
4. Write down what you learned about total eclipse, annular eclipse and a partial eclipse on the lines.
5. Glue the bottom side of your Solar Eclipse Book anywhere in the Sun lesson of your notebooking journal.

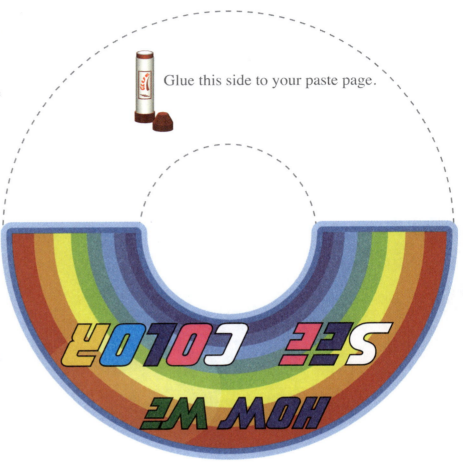

Sun Minibooks: Lesson 2

MERCURY MINIBOOKS

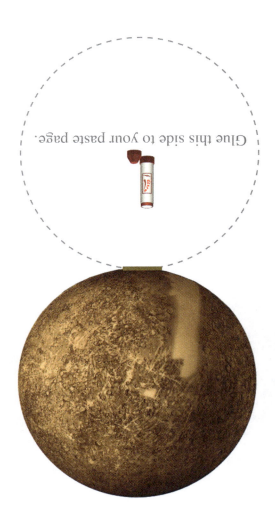

Instructions:

1. Cut out your Mercury Minibooks along the dotted lines. **Do not cut the gold fold lines!** Fold your books in half.
2. Write facts you learned about Mercury on the inside lines.
3. Glue your minibooks to your "Mercury Minibooks" paste page (*NJ p. 49*).

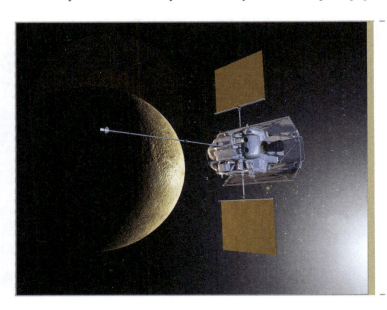

Mercury Minibooks: Lesson 3   A17

VENUS VOLCANO

Instructions:

1. Cut out the Venus volcano.
2. Cut an opening in the top of the volcano along the white dotted line. Be careful not to cut into the volcano. Make sure you cut the entire dotted line. You might use scissors to poke a hole in the white line and then cut the opening with the scissors inserted.
3. Turn the volcano over and dab glue around the outside edges only. **Do not rub glue all over the volcano. Be careful not to glue the opening at the top closed.**
4. Glue your volcano to your "Venus" paste page *(NJ p. 63)*. Be careful not to glue the slit closed.
5. Follow the directions on the next two pages to make the clouds for your volcano.

Venus Volcano: Lesson 4

Instructions:

1. Cut out the clouds (one on this page and four on the next). **Do not cut the gold fold lines!**
2. Fold each cloud along the solid gold line with the words facing outside.
3. Inside each cloud, write the information you've learned about the topic listed on the outside of the cloud.
4. Cut out the strips on this page along the dotted lines.
5. Dab glue on the glue squares (located on the back of each cloud) and affix the clouds to the strips. Let the glue dry.
6. Insert the strips (from smallest to largest) at an angle into the volcano's slit opening.
7. Pull up on each cloud and open to read what you've learned about Venus.

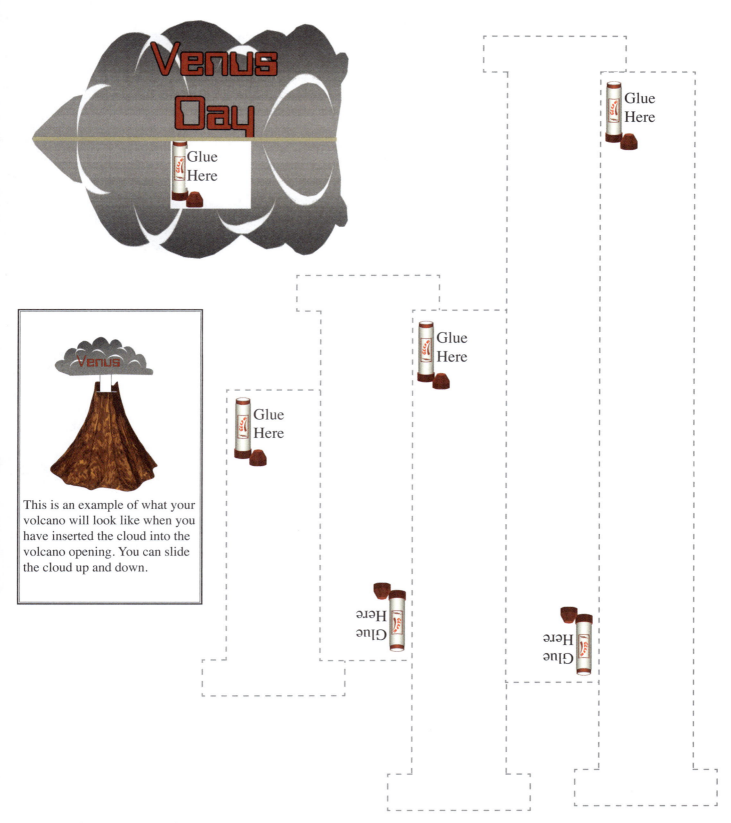

This is an example of what your volcano will look like when you have inserted the cloud into the volcano opening. You can slide the cloud up and down.

Venus Volcano: Lesson 4   A21

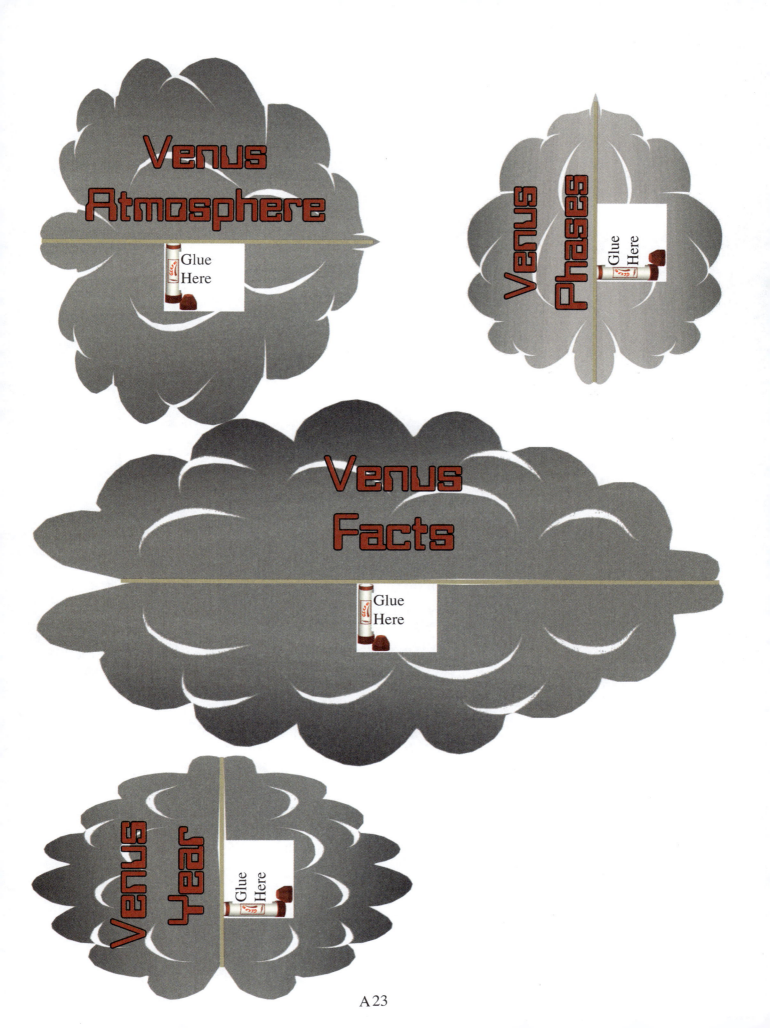

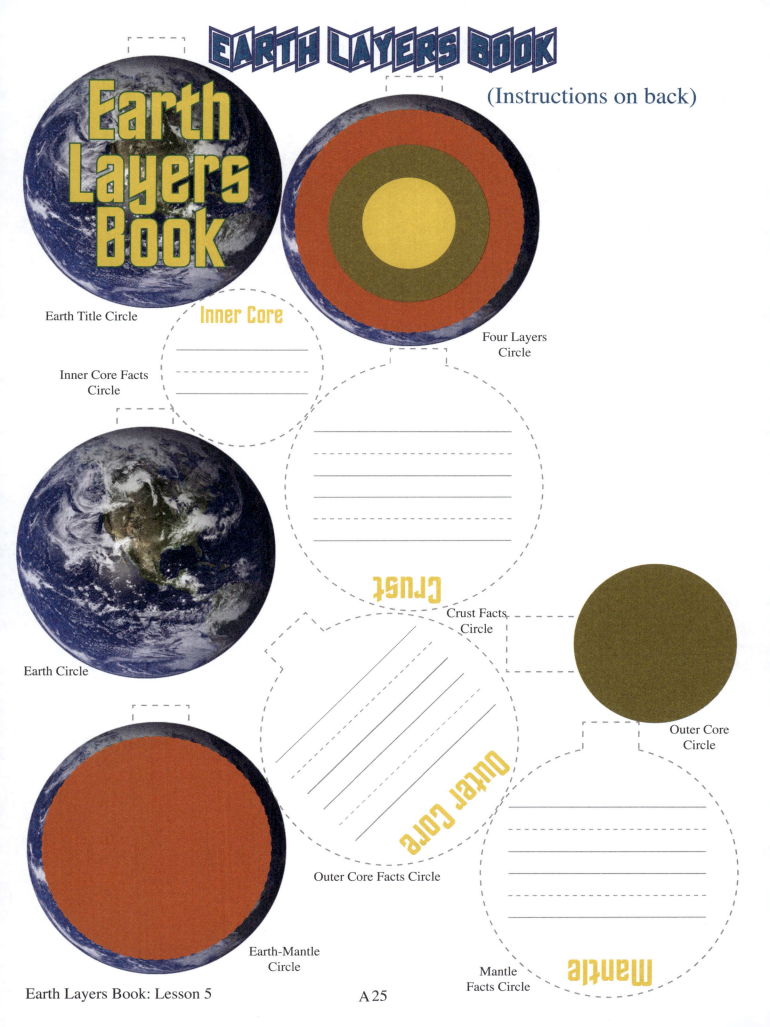

Instructions:

1. Cut out each circle with the tabs attached.
2. Write down the information you've learned on the lines under each topic, including: where the layer is located inside the earth (how far down) and of what materials it is made (Magma, Nickel).
3. Glue the Crust Facts Circle upside down to the Earth Title Circle.
4. Glue the Mantle Facts Circle upside down to the Earth Circle.
5. Glue the Outer Core Facts Circle upside down to the Earth Mantle Circle.
6. Glue the Inner Core Facts Circle upside down to the bottom of the Outer Core Circle.
7. Glue the back of the Four Layers Circle onto the right hand side of your "Earth" Minibooks paste page *(NJ p. 77)* You want to be sure to leave room for your Earth Lift Book.
8. Stack the circles on top of the Four Layers Circle in the correct order with the picture facing up and the facts facing down (Inner Core, Outer Core, Mantle, Crust); and glue them together on the rectangle tabs at the top.
9. Fold the tabs back as you lift each page of your Earth Layers Book. Each layer should display the definition and written information on the tab above the layer.

EARTH LIFT BOOK

Instructions:

1. Cut out the large rectangle along the outer dotted lines. **Do not cut the green fold line! Do not cut the divider lines at this time.**
2. Fold the rectangle along the solid green line so that the words face outward.
3. Cut along the divider dotted lines to create eight flaps. Under each flap, write how each feature listed makes the earth habitable. If you have room, also write what would happen if the earth did not have that perfect feature.
4. Lift the flaps to enjoy reading about the earth.
5. Glue this side of the book onto your "Earth Minibooks" paste page (*NJ p. 77*).

Earth Lift Book: Lesson 5   A27

(Instructions on next page)

The Moon

Phases of the Moon

○ ○ ○ ○ ○
New Crescent Quarter Gibbous Full

How the Moon Lights Up

Place The Moon cover page on top of this layer. Line up at this edge.

Phases

Moon Layered Book: Lesson 6

Instructions:

1. Cut out the four moon pages (two on this page and two on the previous) along the dotted lines.
2. Write or draw the information requested on each page.
3. Stack the moon pages on top of one another beginning with the longest on the bottom and ending with the short "Moon" title page on top.
4. Staple your pages together along the top straight edge above the title "The Moon."
5. Glue your Moon Book onto your "Moon Minibook" paste page *(NJ p. 88)* and lift the pages to enjoy reading what you learned about the moon.

Moon Layered Book: Lesson 6   A31

MARS WHEELS

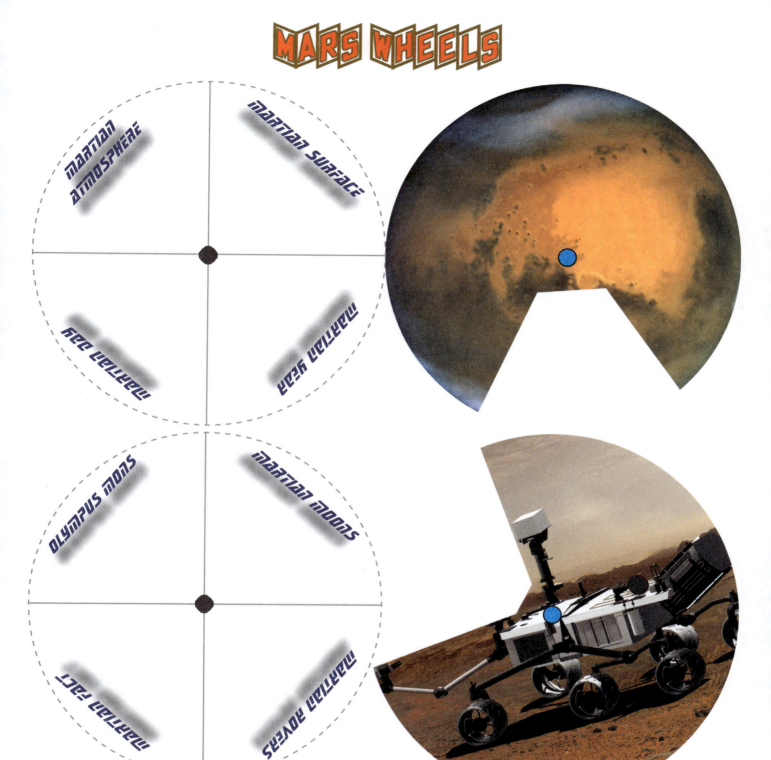

Instructions:

1. Cut out all four circles, along with the cutout in the picture circles.
2. Place each picture circle on top of a fact wheel circle.
3. Secure the top circle to the bottom circle by placing a brass fastener in the center of each circle.
4. Write an interesting fact you learned under the title in each fact wheel.
5. Turn the wheel to reveal the facts about Mars.
6. Glue the bottom of each fact wheel to your "Mars Minibooks" paste page *(NJ p. 100)*.

Mars Wheels: Lesson 7

SPACE ROCKS LAYERED BOOK

Glue the Space Rocks Title Page Here

What are Asteroids?
_____
_____

What is the Asteroid Belt?
_____
_____
_____
_____

ASTEROIDS

Instructions:

1. Write down facts you learned under each title listed on the rectangle pages of your layered book.
2. Cut out all four rectangle pages (Space Rocks, Asteroids, Meteors, Comets).
3. Stack the pages on top of each other with the smallest page on top.
4. Line the pages up at the top with the title of each page showing at the bottom.
5. Staple the pages along the top to secure them together.
6. Glue your layered book onto the "Space Rocks Minibook" paste page *(NJ p. 112)*.
7. Lift the layers to read about space rocks.

Space Rocks Layered Book: Lesson 8

Glue the Asteroids Page Here

What are Meteoroids?

_____

_____

What are Meteors?

_____

_____

What are Meteorites?

_____

_____

What are Shooting Stars?

_____

_____

METEORS

Glue the Meteors Page Here

What are Comets?

_____

_____

Draw a Comet's Orbit

How are Comets Evidence for Creation?

_____

_____

_____

_____

COMETS

Space Rocks Layered Book: Lesson 8

JUPITER SHUTTER BOOK

Glue this side to your paste page.

Instructions:

1. Cut out the Jupiter Shutter Book rectangle. **Do not cut the gold fold lines!**
2. Fold the flaps of the shutter book inward so that the image meets, revealing the complete planet of Jupiter when the book is shut.
3. Fill in facts and information you learned about Jupiter on the lines inside the book.
4. Glue the back down onto your "Jupiter Minibook" paste page *(NJ p. 122)*.

Jupiter Shutter Book: Lesson 9   A39

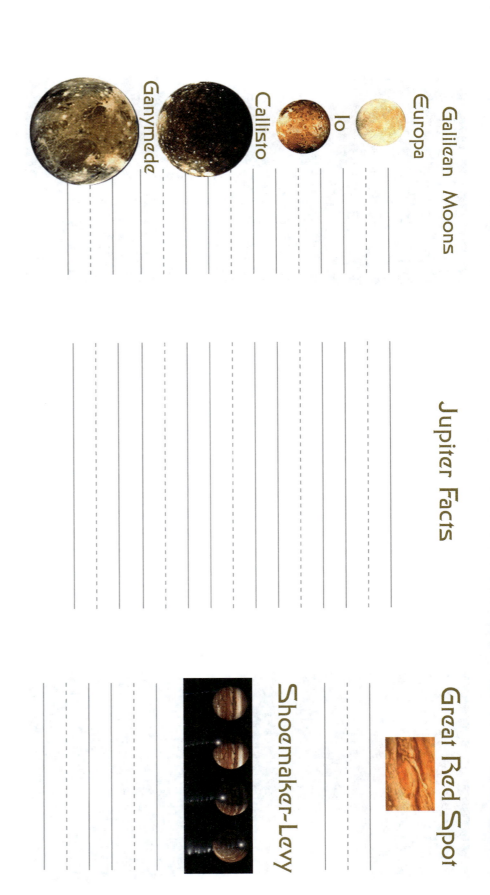

Galilean Moons

- Europa _____
- Io _____
- Callisto _____
- Ganymede _____

Jupiter Facts

_____
_____
_____
_____
_____
_____
_____
_____
_____
_____

Great Red Spot

_____

Shoemaker-Levy

_____
_____
_____

SATURN POCKET BOOK

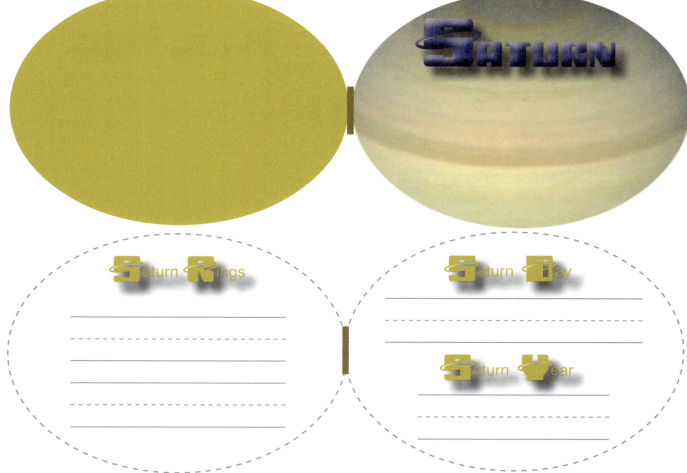

Instructions:
1. Cut out the Saturn Pocket. Create an opening by cutting out the white oval inside Saturn's rings. This is the pocket where you will insert your Saturn Pocket Book.
2. Place glue **on the outside edges only** of your Saturn Pocket.
3. Glue the Saturn Pocket to your "Saturn" paste page being certain not to glue the center opening closed.
4. Cut out the two sets of attached ovals. **Be certain not to cut the gold fold lines!**
5. Fold the ovals at the gold fold lines and place the lined pages inside the cover pages to make a single book.
6. Staple your book on the center line.
7. Write all the facts you learned about Saturn on the lines inside your book.
8. Place your book inside the Saturn Pocket on your "Saturn Minibook" paste page *(NJ p. 133)*.

Saturn Pocket Book: Lesson 10

Saturn Facts

Saturn Facts

Saturn Facts

Saturn Facts

URANUS AND NEPTUNE POP UP BOOKS

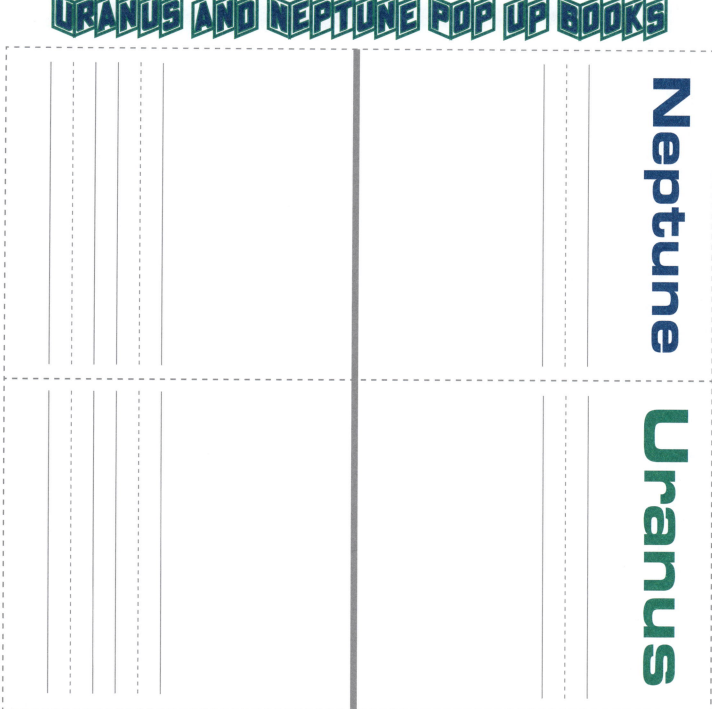

Instructions:

1. Cut out the two rectangles on this page. **Do not cut the grey fold lines!**
2. Write down what you learned about each planet on the lines provided. Fold the rectangles inward along the center grey fold line so that the titles are not visible.
3. Make four small cuts along the dashed lines in the center of each rectangle.
4. Open the paper up and gently pull each pop-up tab forward.
5. Crease the pop-up tabs with your fingers. Close the book to crease the pop-up tabs along the center line so that they are creased outward.
6. Cut out the two colored rectangle covers and the planets on page A45, being careful with Uranus' rings.
7. Fold the covers inward so that the titles are on the outside. Glue them to the outside of your lined rectangles to form the book cover.
8. Glue your planets to the front of the pop-up tabs inside each book.
9. Glue your pop-up books to your "Uranus and Neptune Minibooks" paste page *(NJ p. 145)*.
10. Open your books to see your planets pop up, and enjoy reading all about Uranus and Neptune.

Uranus and Neptune Pop-Up Books: Lesson 11   A 43

These are the covers for your pop-up books. Fold them inward along the center lines and glue them to the outside of your pop-up pages.

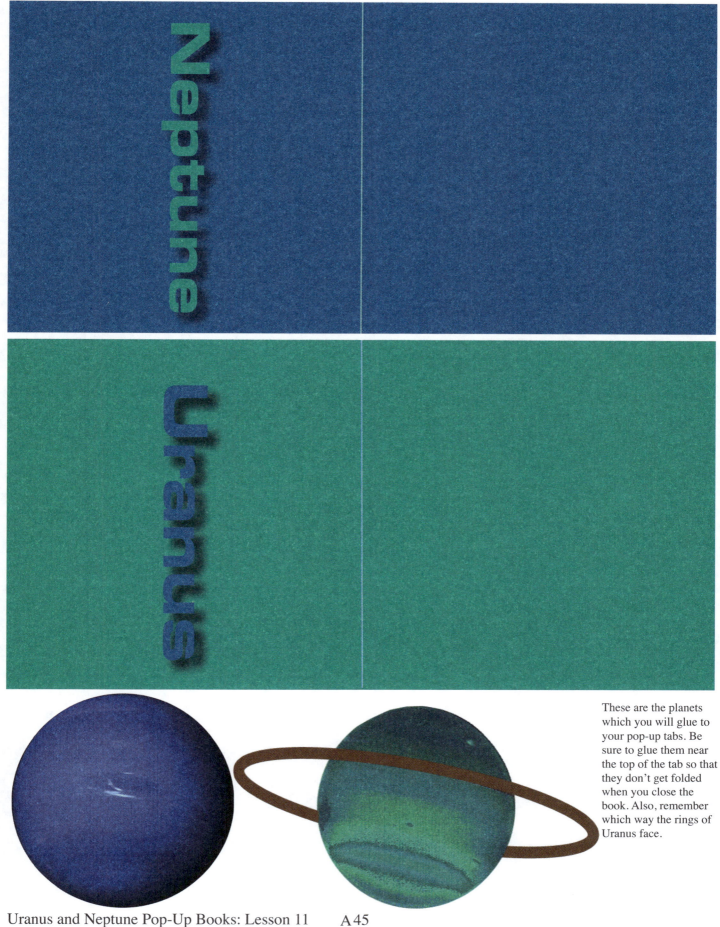

These are the planets which you will glue to your pop-up tabs. Be sure to glue them near the top of the tab so that they don't get folded when you close the book. Also, remember which way the rings of Uranus face.

Uranus and Neptune Pop-Up Books: Lesson 11

Instructions:

1. Cut out the two rectangles on this page along the dotted lines. **Do not cut the grey fold lines! Note the instructions on the back of this page for cutting out the small rectangle.**
2. Fold each rectangle in half along the solid grey lines making sure that the "Is Pluto a Planet?" words are on the outside. This rectangle will be your cover page.
3. Place the lined pages inside the cover page and staple them in the center of the book. This will be the booklet you will use to create your Pluto Debate Book described at the end of Lesson 12 *(T p. 139)* of your astronomy text.
4. Glue this side down onto the "Pluto & the Kuiper belt Minibooks" paste page *(NJ p. 159)*.

Pluto Debate Book: Lesson 12

Instructions:

1. Cut out this small book and fold it in half with title facing outward.
2. Write down what you learned about the Kuiper belt on the inside.
3. Glue this side of the book to your "Pluto & the
4. Kuiper belt Minibooks" paste page *(NJ p. 159)*.

Pluto Debate and Kuiper Belt Book: Lesson 12

STARS & GALAXIES FAN

STARS & GALAXIES FAN

- Which Star Groups are Mentioned in the Bible?
- What is the Milky Way?
- What is Polaris?
- Draw the Four Kinds of Galaxies
  - Spiral
  - Elliptical
  - Lenticular
  - Irregular
- What are Constellations?
- What are Asterisms?

Instructions on next page

Stars and Galaxies Fan: Lesson 13 — A 49

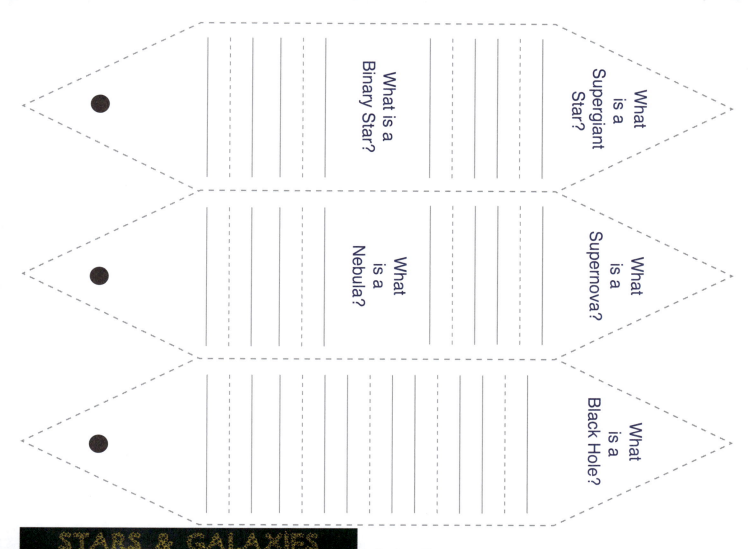

Instructions:

1. Cut out each individual fan sheet (on this page and the previous).
2. Punch a hole in the bottom of each fan sheet on the black dot.
3. Fill in the information requested under each topic.
4. Stack your fan sheets with the Stars and Galaxies Fan sheet on top.
5. Secure the fan sheets at the bottom by inserting a brass fastener into the punch hole.
6. Cut out the pocket to the left.
7. Put glue on the bottom and side edges and paste the pocket onto your "Stars and Galaxies Minibook" paste page *(NJ p. 171)*.
8. Place your Stars & Galaxies Fan in the pocket and remove it when you want to read all about Stars and Galaxies.

Stars and Galaxies Fan: Lesson 13   A51

SPACE TRAVEL LIFT AND LEARN

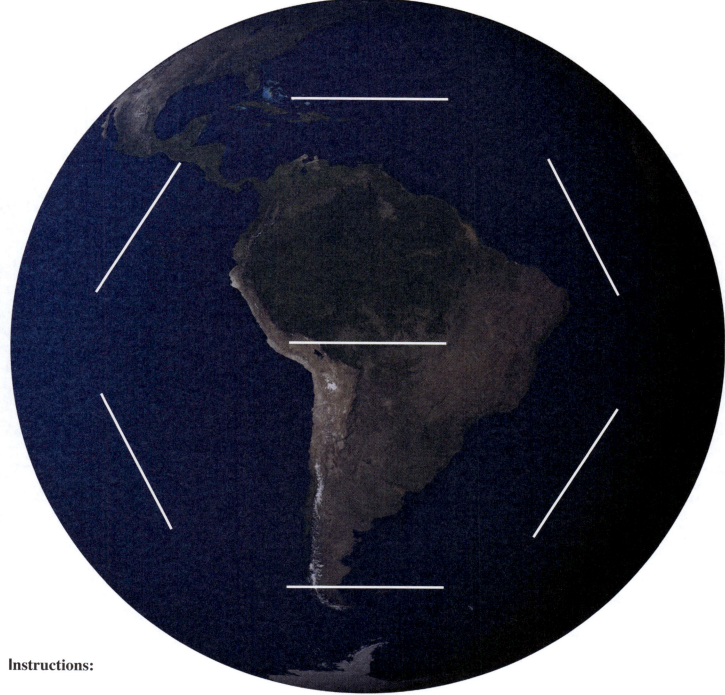

Instructions:

1. Cut out the earth image above. Make small openings in the earth along the white lines.
2. Turn the earth over and place glue around the outer edges. **Be certain not to glue the white openings closed!**
3. Glue the earth image to your "Space Travel Minibook" paste page *(NJ p. 185)*.
4. Cut out each space travel shape (on the next two pages).
5. Answer the questions related to each picture.
6. Insert all the shapes except the astronaut into the openings on the earth and slide them down.
7. Insert the astronaut into the middle and fold him face down at the opening. Then fold him upward at the bottom of his boots to make him pop out and wave to you when you open your notebooking journal.
8. Pull up each shape to learn about space travel.

Space Travel Lift and Learn: Lesson 14

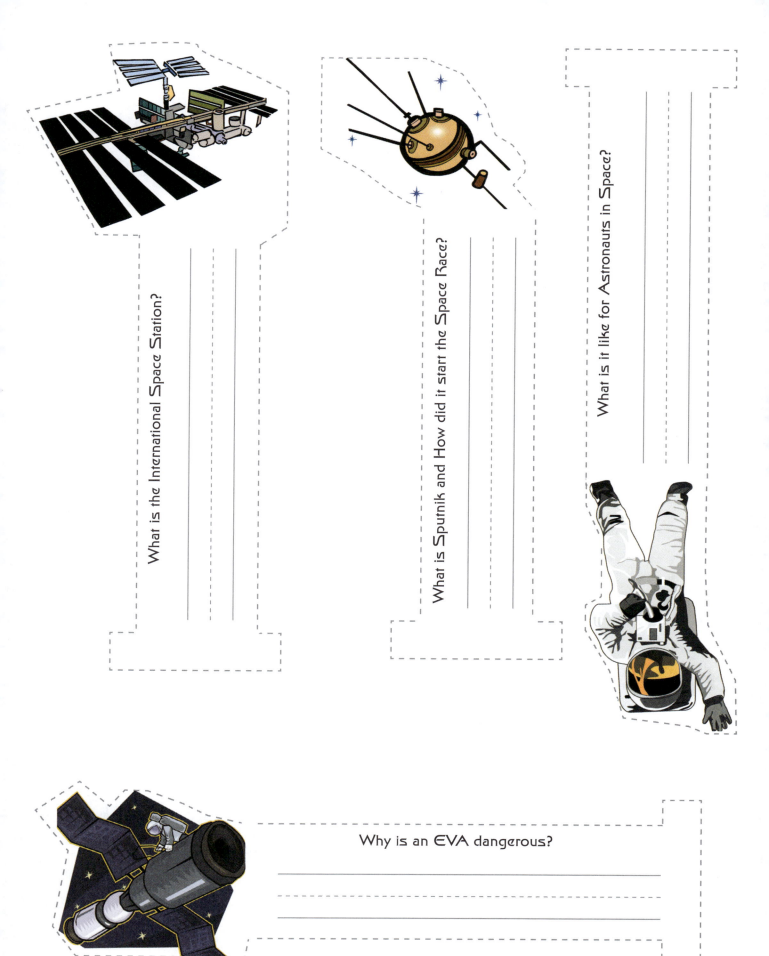

What is the International Space Station?

What is Sputnik and How did it start the Space Race?

What is it like for Astronauts in Space?

Why is an EVA dangerous?

Space Travel Lift and Learn: Lesson 14

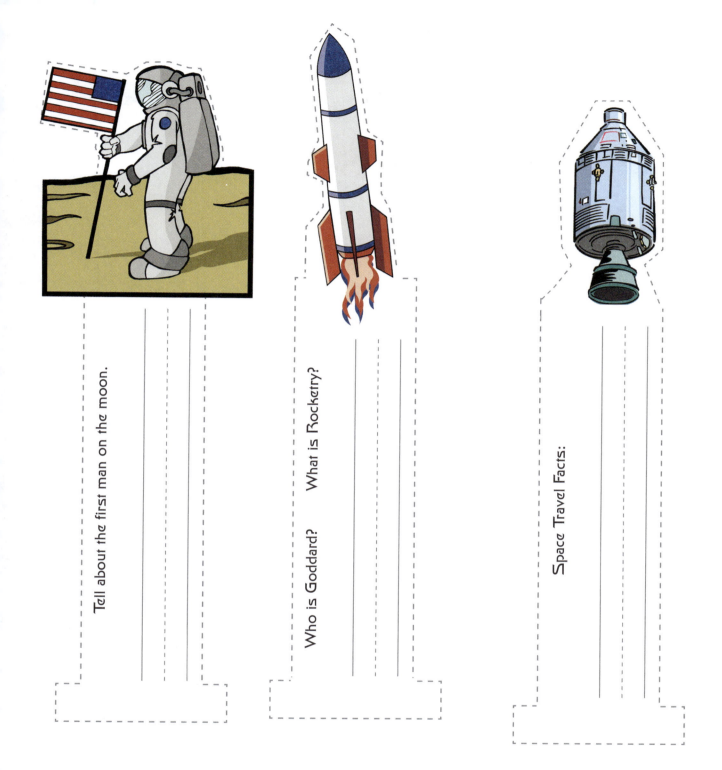

Space Travel Lift and Learn: Lesson 14

SOLAR SYSTEM REVIEW WHEEL

Instructions:

1. Cut out the three circles on this page and the satellite strips on the next page.
2. Write a fact you learned about each planet on the satellite strips. For example, you might write: *"Called the Red Planet"* on the strip below the picture of Mars. **Write the fact on the strip as close to the planet as possible, because the bottom half of Mercury, Venus and Earth will be hidden behind the sun when the wheel is assembled.**
3. Punch holes through all the black dots using a hole puncher. You will need to fold down one edge of the circle to get the hole puncher to the center black dot.
4. Place a brass fastener through the hole in the center of Circle 1.
5. Place all the planets behind Circle 1 by placing the fastener through the holes on each strip. Mercury will be the first planet and Neptune will be the last.
6. Place Circle 2 behind the planets by placing the fastener through the circle's hole.
7. Bend down the prongs of your brass fastener.
8. Glue the Sun Circle to the front of Circle 1, on top of the metal fastener head.
9. Glue the back of Circle 2 to your "Planets Review Minibook" paste page *(NJ p. 186)*.
10. When the satellites are fanned out, your wheel will look like the small picture below.
11. Rotate the satellites so that they are above the sun before you close your book.

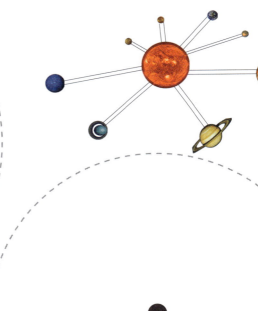

Circle 2

Circle 1

Solar System Review Wheel

Solar System Review Wheel A61